U0146932

"十一五"国家重点图书出版规划项目

21世纪先进制造技术丛书

Haidou Wang
Binshi Xu
Jiajun Liu

Micro and Nano Sulfide Solid Lubrication

（微纳米硫系固体润滑）

With 367 figures

 Science Press
Beijing

 Springer

Authors

Haidou Wang
National Key Laboratory for Remanufacturing
Academy of Armored Forces Engineering
Beijing, 100072, China
Email: wanghaidou@tsinghua.org.cn

Binshi Xu
National Key Laboratory for Remanufacturing
Academy of Armored Forces Engineering
Beijing, 100072, China
Email: xubinshi@vip.sina.com

Jiajun Liu
Department of Mechanical Engineering
Tsinghua University
Beijing, 100084, China
Email: jiajun7458@yahoo.com.cn

ISBN 978-7-03-031785-8
Science Press Beijing

ISBN 978-3-642-23101-8
Springer Heidelberg Dordrecht London New York

Preface

In order to cope with the fast development of modern industries, the lubrication of machinery equipment has attracted extensive attention. Although the liquid lubrication is widely used, more and more machine parts are working in the increasingly severe conditions, like high temperature, high load, ultralow temperatures, ultrahigh vacuums, strong radiation, etc. For these working conditions, the commonly used lubricating oils and greases are hardly applicable, so that the solid lubrication is necessary as the supplement for liquid lubrication.

The solid lubrication technique was first applied to the aerospace and military industries, after that, it was gradually propagated to the high-tech and newly developed departments of industry. Its development enlarged the application fields of the traditional lubricating oils and greases, counteracted their shortcomings, and the most important result is the emergence of an increasingly large amount of new lubricating techniques using solid lubricants, which are either new materials, or new techniques, exhibiting more important meanings. The practice has proven that the proper adoption of advanced lubrication materials and techniques can effectively prolong the service life of machinery working under the harsh conditions. Furthermore, saving energy and raw materials is the general and urgent requirement today. The rational application for lubricating materials and techniques are even more important economically and socially.

This book aims to introduce the advanced solid lubrication techniques of sulfur system to a broad range of readers. It presents the recent research successes of the authors. Based on this, the authors, depending on their profound theoretical basis and experiences in the interdisciplinary field covering tribology, materials science, surface engineering, chemistry, physics and so on, explained the solid lubrication of materials and techniques of sulfur system comprehensively and systematically. They involve microstructures, properties, preparation methods, applications and mechanisms of the solid lubrication materials. Hope this book can be useful to those readers who are interested in the research and application for solid lubrication.

The main contents of this book are selected from the authors' basic research work in the recent twenty years. However, the solid lubrication is a new area of tribology and lubricating materials; a special and interdisciplinary subject, most phenomena and mechanisms are still in their early stages. What we have discussed in this book may be incomplete or not one hundred percent correct, for this reason, we welcome different opinions or comments from readers.

The authors would like to thank Dr. Lina Zhu and Dr. Jiajie Kang for their supports. This book wouldn't be finished without their help.

Finally, we are grateful to the National Natural Science Foundation of China (50575225), Beijing Science Foundation (3072011) and Equipment Pre-investigation Foundation (9140A27030107OC85) for their supports towards our research works.

Haidou Wang   Binshi Xu   Jiajun Liu

July 2011

Contents

Chapter 1
Solid Lubrication Materials

Solid lubrication is an important component of lubrication. It overcomes some inherent drawbacks for liquid lubrication. Solid lubrication can be adopted in very harsh conditions such as high temperature, heavy load, ultralow temperature, ultrahigh vacuum, strong oxidation, and intense radiation, etc. In above conditions, liquid lubrication will lose its lubricating function. Solid lubrication materials are divided into four categories, i.e., soft metals, metal compounds, inorganic and organic materials. They have different lubricating mechanisms and can be applied to different situations.

1.1 Overview of Solid Lubrication

1.1.1 Introduction

Friction leads to the loss of a large quantity of mechanical energy, while wear is an important reason of the failure of machinery parts. According to incomplete statistics, about 1/3 of the energy was consumed by friction; and about 80% failure of machine parts was caused by wear. For a highly industrialized country, the annual economic loss caused by friction and wear almost accounts for 2% GDP value of the national economy. Therefore, the research on friction and wear is a long term subject, which has important social and economic benefits.

From time immemorial, friction and wear have been accompanying human life and production, and the efforts that people try to control friction and reduce wear have never been stopped. With the development of industry, especially in modern industry and technology, ever-increasing demands are put forward for friction and wear in the harsh working environment such as high speed, heavy load, high vacuum, high and low temperature, nuclear radiation. Such a background provided strong impetus for the development of the interdisciplinary subject of industrial tribology. In recent years, the application of surface engineering techniques for improving the friction-reducing and wear-resistance of materials has played a very important role in elevating quality of products, decreasing cost and saving energy. Surface engineering technology can modify the materials surface of machine parts, greatly improve their performance and significantly extend their service life through preparing various coatings and films. Meanwhile, it can be also used for repairing the failed machine parts. So, surface engineering technology will be one of the key technologies in manufacturing and remanufacturing of mechanical equipment, which will master the industrial development in the 21st century. The rapid development of surface engineering provided a very good opportunity for the R&D of solid lubrication materials, in other words, the great progress of solid lubrication in recent years is promoted by the development of surface engineering technology.

1.1.2   Adhesive Wear and Scuffing of Metals and Methods of Prevention

1.1.2.1   Essence of adhesive wear and scuffing

When two surfaces come into contact, load is actually born by many asperities; the real contact area of rough surfaces is usually far less than the nominal contact area. Therefore, the local high stress is easily generated at the asperity peaks. When the surfaces are sliding with each other, the local melting caused by friction heat will lead to the adhering or welding between asperities, which can induce the macro adhesion and scuffing, or even seizure of friction surfaces. Adhesive wear is a common wear mode of the friction-pair in machinery equipment. Due to high surface energy and large chemical activity, clean metal surfaces are prone to adhesion, thus the friction coefficient is large, and the wear is severe. Under oil lubrication condition, adhesion can occur if the oil film is damaged; otherwise, the friction coefficient and wear can maintain a relatively low level. According to the degree of adhesion, adhesive wear can be divided into four categories. When soft metal material transfers to hard metal surface, and the transfer film is very thin, it is known as smearing. When shearing occurs within the subsurface layer of soft metal, and hard surface is also scratched, it is known as scratching. When shearing occurs in the deep layer of substrate metal, it is known as scuffing. Cold welding caused by plastic deformation and molecular absorption is known as the first type of scuffing, while hot welding caused by the rise of surface temperature is known as the second type of scuffing. When external force can not overcome the bonding strength of interface, the relative movement is forced to be stopped, which is known as seizure.

Under low-load condition, oxidation wear will happen firstly on the friction surface, the generated oxide film can prevent the occurrence of adhesion; when the load is increased, the oxide film will be destroyed, the lubrication effect is lost, and the adhesive wear will be present. The existence of liquid or solid lubricating films can effectively prevent the development of adhesive wear. There are many factors influencing the adhesive wear, in general, including two aspects: one is the own characteristics of the friction-pair materials, like intersolubility of metals, crystal structure, microstructure, and hardness, etc.; the other is working conditions, such as load, sliding speed, working temperature, and lubricating condition, etc.

Scuffing is a serious wear mode, it usually appears in the sliding wear; its definition is defined as "severe wear with characteristic of forming the partial weld spot between sliding surfaces". It indicates that it was generally accepted that the essence of scuffing is belonging to adhesive wear, but the subsequent researches proved that adhesive wear is only one of the mechanisms of scuffing and mostly occur in the case of dry friction. Under oil lubrication condition, the scuffing mechanism is mainly strain fatigue wear present in the form of delamination. When load and sliding speed increase to a certain extent, the boundary lubricating film will be damaged, or it is insufficient to form boundary lubricating film; the scuffing mechanism will convert from fatigue wear to adhesive wear.

1.1.2.2   Surface modification

In order to improve the wear-resistance of machine parts, a wide range of surface modification technologies are employed, such as surface quenching, surface chemical heat treatment, electric plating, laser modification, thermal spraying, superhard film depositing, ion implantation, etc. Surface modification can change the structures and properties of friction-pair surfaces, weaken the mutual solubility between materials, increase hardness and strength of surface, as a result, the deformation of surface can be reduced, the adhesion and scuffing of friction-pair are mitigated or avoided. But for surface hardening materials, when the center hardness is high, and hardened layer is thick, the shallow flaking will happen; when the center hardness is low, and hardened layer is thin, the whole hardened layer will peel off. The mechanical conditions leading to the two kinds of flaking are shown in Figure 1.1. For stress fatigue wear, crack initiation is the dominant process; thus the higher the material hardness, the more difficult the crack initiation, and the longer the fatigue wear life. For the strain fatigue wear, the propagation of fatigue crack is the dominant factor to determine the fatigue life. The increase of hardness must lead to decrease of the material toughness and accelerating the propagation rate of cracks, which makes scuffing happen easily. Therefore, for resisting adhesive wear and fatigue wear, high hardness of material surface is not always necessary. Many researches have proved that the high carbon steel in the normalizing state possesses better anti-scuffing performance than that in the quenching state. Therefore, in the case for improving the anti-scuffing property of friction surface as the main purpose, the modification techniques of establishing a solid lubrication coating (soft coating) on the hard substrate should be adopted, but not to increase the surface hardness blindly.

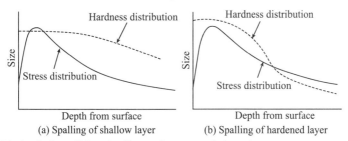

Fig.1.1   Mechanical conditions leading to the contact fatigue damage

Figure 1.2 shows the principle scheme how the solid lubrication coating can protect the material surface from strain fatigue wear (delamination).

1.1.2.3   Liquid lubrication and additive

Under the condition of hydrodynamic or elastohydrodynamic lubrication, the lubricating oil film will become thinner if its viscosity decreases, or the load is increased or speed is decreased. When the thickness of oil film is less than a certain value, asperities come into contact; the friction-pair will enter the state of boundary lubrication. In this case, the active elements in the oil can react with the metal surface to generate chemical reaction film under the effect of friction heat. Figure 1.3 shows the

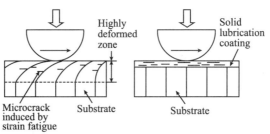

Fig.1.2 Principle scheme how the solid lubrication coating can protect the material surface from strain fatigue wear (delamination)

schematic diagram of the formation of FeS reaction film. The thickness of FeS reaction film is very small not more than $0.2\mu m$, but the bonding strength between the film and substrate is enough high. In addition, the reaction is completely non-reversible and has greater stability compared with physical adsorption and chemical adsorption; therefore, adding various additives to base oil is an important method to achieve excellent lubrication effect. In a large number of wear-resistant and extreme pressure additives, sulfur and organic sulfide are a very important kind of additives. Batchelor has pointed out that when the hexadecane containing 0.75wt% sulfur is deposited on the fresh steel surface, a 75Å thick sulfide film can be generated within 3×10^{-2}s, and it is enough to make the adhesion of surface decrease to a safe range. As the formation time of ion sulfide film is short, when the lubricating oil contains sulfur, the scuffing is hard to occur even in very poor lubrication condition. At $100\text{---}170°C$, the reaction of organic disulfide and iron on the friction surface is as follows:

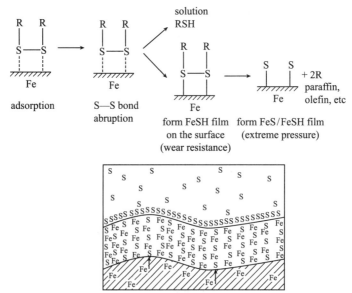

Fig.1.3 Schematic diagram of the formation of FeS reaction film by sulfur and iron

Disulfide is absorbed on the metal surface, and S—S bond disconnects to form mercaptan iron film, which can effectively protect the surface if the friction condition is not very serious. Under extreme pressure and heavy load, C—S bond disconnects to form iron sulfide, which can protect the surface as well.

Organic sulfide as boundary lubricating additive can play a good role only with the existence of air (oxygen). The researches of Baldwin have shown that under the extreme pressure condition, the balance product on the friction surface is $FeSO_4$, but it usually can not be obtained, as a result, FeS is present. For each kind of sulfur-containing additive, the formed boundary film always contains iron sulfide and iron oxide, among them iron oxide plays the role of catalysis to make the active sulfur atoms dissociated from sulfide and reacted with iron on the friction surface. Sulfur atoms can be replaced easily by oxygen atoms when the iron sulfide is in oxygen atmosphere; however, the sulfide can not play an effective role without the existence of oxygen.

Although liquid lubrication has a wide range of application, with the development of industry and technology, more and more machine parts are applied in the very harsh conditions, such as high temperature, heavy load, ultralow temperature, ultrahigh vacuum, strong oxidation, and intense radiation, etc. the general liquid and grease lubrications have lost their function in above conditions. To this end, solid lubrication can be adopted only.

1.1.3   Solid Lubrication

1.1.3.1   Definition and significance

According to the tribological terminology formulated by the Organization of Economic Cooperation and Development (OECD), the definition of solid lubrication is "use of any solid powders or films, which can protect the surfaces in relative motion from damage and reduce the friction and wear". In the course of solid lubrication, the physical and chemical reactions will happen between the solid lubricant and surrounding medium to reduce friction and wear [1].

Solid lubrication technology was first applied in military industry, and then applied to some high-tech fields such as artificial satellite, spacecraft and electronic industry. It had solved many problems, which could not be solved by liquid lubrication, and had been successfully applied in a variety of special conditions. Therefore, more and more attentions have been paid to solid lubrication technology. On the other hand, the global energy resources have been getting tense at present, thus the voices of that solid lubrication will replace liquid lubrication get more and more rising. So that the theoretical studies on the lubrication mechanism are increasing day by day; the effectiveness of solving the lubrication problems by solid lubricant is increasingly significant.

The application of solid lubricants has already a long history. Graphite, molybdenum disulfide, plumbum salt, soft metal powders and other solid lubricants have been applied well in industries, for example the polytetrafluoroethylene powder has been applied successfully in lubricating oil and grease as additive.

1.1.3.2   Role of solid lubricant

The solid lubricants overcome some inherent drawbacks for liquid lubricant. For instance, lubricating oil and grease are easy to evaporate, and the vapor pressure is high; thus they can not be used for a long time in the high vacuum of more than 10^{-1}Pa. However, in the outer space at the height of 1000 km, the vacuum degree is very high; the absolute pressure can achieve 10^{-3}—10^{-2}Pa. Therefore, artificial satellite needs the solid lubricant with low vapor pressure.

For the rocket carrying satellite, the kerosene and liquid hydrogen are used for propellant; liquid oxygen is used for oxidant. As the liquid hydrogen and oxygen are ultralow temperature fluid, with the boiling points of $-253°$C and $-183°$C respectively, the lubricating oil can not be used when the pressured transmission from tank to the turbopump support shaft of combustion chamber is conducted, especially once the liquid oxygen and lubricating oil contact, explosion will happen. Therefore, the bearings on the turbopump can be lubricated only by solid lubricant.

When liquid lubricant bears heavy load, the oil film will be damaged and lose the lubricating function at high temperature, while solid lubricant possesses high bearing capacity and can resist high temperature.

1.1.3.3 Application conditions of solid lubricants

1. The conditions, where solid lubricant can replace lubricating oil and grease

Under the following conditions, solid lubricants can replace lubricating oil and grease to lubricate friction surface.

(1) Solid lubricants can be used in various special conditions such as high temperature, low temperature, vacuum and heavy load, etc. The common lubricating oil and grease can not meet their requirements.

(2) Solid lubricants can be used in conditions where lubricating oil and grease are easily polluted or washed away by other liquids (such as water, seawater). They can also be used in the moist environment and solid impurities (such as silt, dust) existing conditions.

(3) Solid lubricants can be used in the conditions where lubricating oil and grease are difficult to be supplied to some components and friction-pairs, it is difficult to approach or load and unload during installations, and it is difficult to operate maintenance periodically.

2. To improve the properties of lubricating oil and grease

Solid lubricant can be added to lubricating oil and grease with following purposes:

(1) To improve the load-carrying capacity of lubricating oil and grease;

(2) To improve the ageing properties of lubricating oil and grease;

(3) To improve the high-temperature properties of lubricating oil and grease;

(4) To make the lubricating oil and grease form tribological polymer film.

3. Harsh running conditions

(1) Lubrication under a wide range of temperature.

The service temperature of lubricating oil and grease is about -60—$350°$C, while solid lubricant can be used under the temperature ranging from $-270°$C to above $1000°$C.

The solid lubrication under ultra-low temperature is the key of ultra-low temperature technology. Solid lubricants like PTFE, and plumbum, etc. still possess lubricating properties under this temperature.

The ceiling service temperatures of high molecular material polyimide and plumbum oxide are 350°C and 650°C respectively, while that of the mixture of calcium fluoride and barium fluoride is 820°C. The operating temperature of hot rolling of steel products can reach more than 1200°C. In this condition, graphite, glass, and all kinds of soft metal films can be used as effective lubricant.

(2) Lubrication under a wide range of speed.

A variety of solid lubricating films can be used under a wide range of sliding speed. For instance, the motion of machine tool guideways belongs to low speed motion, the lubricating oil containing solid lubricant can reduce crawling, and the solid lubrication dry film made from high molecular material by spraying method can reduce wear. The soft metal plumbum film can be applied to sliding surfaces of low speed motion. The bearings applied in the condition of low speed and heavy load can be made from sliding materials in the form of mosaic structure. For the bearings used in the high speed condition, when titanium carbide film with thickness of 2—5μm is deposited on their surface, the wear is slight even if they are operating at a speed of 24000r/min for 25000h.

(3) Lubrication under heavy load.

The oil films of common lubricating oil and grease can sustain only relatively low loads. Once the load exceeded the limited value, the oil film will be damaged, and the seizure may occur. However, solid lubrication film can bear the average load over 10^8Pa. For instance, MoS_2 film with thickness of 2.5μm can bear a contact pressure of 2800MPa and operate at a high speed of 40m/s. PTFE film can bear a pressure of 10^9Pa, while the load-carrying capacities of metal-based composite materials are even higher. For the pressure working of metals such as rolling, extruding, or stamping, the load is very high. In this condition, oil-based or water-based lubricant containing solid lubricant or solid lubrication dry film can be used.

(4) Lubrication under vacuum.

Under vacuum or high vacuum conditions, common lubricating oil and grease have relatively large evaporability, which easily destroys the vacuum condition, and affects the properties of other components. In this condition, metal-based and high molecular polymer-based composite materials are commonly used.

(5) Lubrication under radiation.

Under the radiation condition, general liquid lubricants will be polymerized or decomposed, and lose their lubrication properties. Solid lubricants have better radiation resistance. For instance, the radiation resistance of metal-based composite materials is more than 10^8Gy; no detectable change happens after graphite is radiated by strong neutrons of 10^{20}/cm^3. Under the radiation condition, layered solid lubrication materials, polymer-based and metal-based composite materials can be used.

(6) Lubrication for conducting sliding surfaces.

For the friction between conducting sliding surfaces, such as electric brush, conductive slider, solar slip rings on the man-made satellite working in vacuum, and sliding electrical contact, carbon-graphite, metal (Ag)-based composite materials, or composite material composed of metal and solid lubricant can be used.

4. Formidable environment conditions

The transmission parts of transport machinery, construction machinery, metallurgy machinery, and mining machinery, etc. are working mostly in the formidable environment, such as dust, silt, high temperature and moisture, etc., solid lubricants can be used in this case.

5. Clean environment conditions

The transmission parts of machinery in the electron, textile, food, medicine, paper making, and printing industries, etc. should avoid pollution and require clean environment. They can be lubricated with solid lubricants.

6. Maintenance-free conditions

Some transmission parts do not need maintenance, or the maintenance frequency for some transmission parts should be reduced to save expenditure. In these conditions, it is rational and convenient to use solid lubricant.

(1) Unmanned and maintenance-free conditions.

For the supporting of large and medium bridges and transmission parts of large and heavy-duty equipment, it is difficult to maintain. In order to reduce maintenance expense, prolong the service life of machines, and extend the effective operating period under service-free conditions, solid lubricant can be used. For long-term storage and easy maintenance firearms and instantaneous products, solid lubricating dry films can be used.

(2) Frequent dissembling and maintenance-free condition.

When the fasteners, screws, and nuts, etc. are covered by solid lubricating dry films, they are easily loaded and unloaded; meanwhile, the fretting wear of fasteners can also be prevented.

1.1.3.4   The employed methods for solid lubricant

1. Usage as integral parts

As some engineering plastics, such as polytetrafluoroethylene, polyacetal, polyoxymethylene, polycarbonate, polyamide, polysulfone, polyimide, chlorinated polyether, polyphenylene sulfide, and poly p-phenylene dicarboxylic acid vinegar have low friction coefficient, good formation ability and chemical stability, excellent electric insulating property and strong shock resistance, they can be made into integral parts. Their comprehensive property will be better if these plastics are strengthened by glass fiber, metal fiber, graphite fiber and boron fiber, etc. These engineering plastics are widely applied to gears, bearings, cams, rolling bearing cages, etc.

2. Usage as composite materials or combination of materials

As the physical and chemical properties and shapes of solid lubricating materials are different and immiscible, the composite materials with better performances can be obtained by combining these solid lubricating materials.

3. Usage as solid lubricating powders

The load-carrying capacity and boundary lubricating conditions can be improved when proper quantities of solid lubrication powders (such as MoS_2) are added to lubricating oil and grease.

4. Usage as surface layers

Solid lubricants can be applied to form films or coatings with a certain self-lubricating properties on the friction surface. This is a commonly used method.

1.1.3.5   Species of solid lubricant

There are a number of solid lubricants, and they have different lubrication mechanisms. Classified by basic raw material, they can be divided into soft metals, metal compounds, inorganic and organic materials, etc. Many substances can be used as solid lubricants, among them the commonly used materials are layered structure materials such as graphite and molybdenum disulfide, soft metals such as plumbum and argentine, and high molecular materials such as PTFE and nylon.

1.2   Soft Metal

Many soft metals such as Pb, Sn, Zn, In, Au and Ag have better lubricating effects and can be used as solid lubricant under conditions of radiation, vacuum, high temperature, low temperature and heavy load. Commonly, the soft metals can be made into alloy materials or coated on friction surfaces by electric plating, or other techniques to form solid lubricating film.

Soft metal solid lubricating materials used as solid lubricants are base on their low shearing strength and easy intergranular slip. The soft metals with certain strength and toughness can firmly adhere to the substrate surface and play the role of friction-reduction and lubrication.

1.2.1   Crystal Structure

Soft metals, like plumbum has face-centered cubic lattice, as shown in Figure 1.4. As their crystals are anisotropic, these soft metals can not possess high load-carrying capacity; the plus load can only be born by substrate.

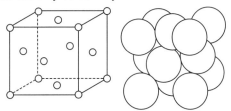

Fig.1.4   The crystal structure of plumbum

As crystals do not have directivity, the lubrication properties of soft metals are similar to those of high-viscosity fluid. For instance, plumbum film has self-repairing property at low friction speeds; therefore, it is effective when plumbum film is used as lubricant for the low-speed operation machinery.

Another advantage of face-centered cubic crystals is that they do not have low temperature brittleness, in other words, they do not lose lubricating properties under low temperatures. It was also found that it can be used as lubricant under ultra-low temperature. Therefore, soft metal solid lubrication films have wide application temperatures ranging from ultra-low temperature to a certain degree of high temperature.

1.2.2   Physical and Chemical Properties

Mechanical properties of commonly used soft metals are shown in Table 1.1.

Table 1.1 Basic properties of soft metal solid lubricating materials

Material	Element symbol	Density/(g/cm³)	Melting point/°C	Crystal type	Coefficient of linear expansion /(10⁻⁶/°C)	Tensile strength /(MPa×9.8)	Extensibility/%	Rate of contraction /%	Hardness HB	Critical shearing stress /(MPa×9.8)
Aurum	Au	19.32	1063	Face-centered cubic	14.2	14	50	90	18	0.092
Argentine	Ag	10.49	960.8	Face-centered cubic	19.7	18	50	90	25	0.060
Plumbum	Pb	11.34	327.3	Face-centered cubic	29.3	1.8	45	100	4	
Zinc	Zn	7.13	419.5	Close-packed hexagonal	39.5	15	20	70	30	0.030
Tin	Sn	7.3	231.9	< 10°C Diamond-type > 16°C Cubic face-centered	23	5.2	40	100	5	0.013—0.019
Indium	In	7.31	156.6	Face-centered cubic	33				3	
Cadmium	Cd	8.65	321		31	6.4	20	50	20	

1.2.3 Lubrication Mechanism

The solid lubrication effect of soft metals is due to its low shearing strength. In the process of friction, soft metal can form transfer film on the counterface and make the friction occur between the transfer film and soft metal; therefore, the friction coefficient and wear can be reduced.

The higher the purity of soft metal, the lower the shearing strength, and the easier the slip within the soft metal. Therefore, it has self-repairing property. Soft metal has the similar lubricating property to high-viscosity fluid, once the damage of lubricating film is present, it can play the role of self-repairing through the action of applied load to recover the lubricating property.

1.3 Metal Compounds

Lots of metal compounds can be used as solid lubricants, such as metal oxide, halide, sulfide, selenide, borate, phosphate, and organic salt.

Metal oxide includes plumbum oxide, plumbum tetraoxide, antimony oxide, antimony trioxide, chromic oxide, titanium oxide, zirconium oxide, iron oxide, magnetic iron ore, and aluminum oxide, etc. The lubricating effect can be better when these oxides are used at high temperature.

Metal halid includes calcium fluoride, barium fluoride, lithium fluoride, cerium fluoride, lanthanum fluoride, boron fluoride, cadmium chloride, cobalt chloride, chromium chloride, nickel chloride, iron chloride, boron chloride, copper bromide, calcium iodide, plumbum iodide, cadmium iodide, argentine iodide, and mercuric iodide, etc.

Metal sulfide includes iron sulfide, molybdenum disulfide, tungsten disulfide, zinc sulfide, and plumbum sulfide, etc.

Metal selenide includes tungsten diselenide, molybdenum diselenide and niobium diselenide, etc.

Metal borate includes kalium borate ($K_2O \cdot 3B_2O_3 \cdot 2H_2O$) and sodium borate, etc.

Metal sulfate includes argentine sulfate and lithium sulfate, etc.

Metal organic salt is composed of various metal fatty soaps, such as calcium soap, sodium soap, magnesium soap and aluminum soap. They all have relatively good lubricating properties.

Several kinds of chalcogenide solid lubrication materials are mainly introduced as follows.

1.3.1 FeS

Sulfur is regarded as one of harmful elements in the irons and steels; its content is strictly controlled. However, people have found that the wear-resistance of workpiece can be improved significantly when its surface contains sulfur, or the compounds of sulfur are added into lubricating oil; especially it is effective to prevent the seizure and scuffing of workpieces. In the 1920s, a scientist named Zaitsev of former Soviet Union analyzed this phenomenon in detail through plenty of experiments. He considered that once the sulfur was present on the surface during friction, a thin FeS film would be formed. This film with excellent friction-reducing behavior increased the resistance to scuffing and seizure.

1.3.1.1 Crystal structure of FeS

Figure 1.5 shows the schematic diagram of crystal structure of FeS. The structure of FeS is similar to that of graphite and molybdenum disulfide, namely close-packed hexagonal lattice. The lattice constants are $a = 0.597$nm, $c = 1.174$nm. FeS exhibits squamose appearance. It has low resistance to deformation and shearing strength; it easily slips along the plane of $\{0001\}$.

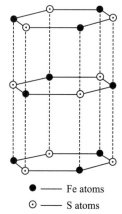

● —— Fe atoms

⊙ —— S atoms

Fig.1.5 Schematic diagram of crystal structure of FeS

1.3.1.2 Properties of FeS

FeS is a kind of non-metal material, and has good cold-welding resistance. FeS has a high melting point reaching 1100°C. It has low hardness (HV 100), good plastic flow performance and excellent lubricating property.

FeS possesses good anti-oxidation property; the oxidation speed of dry iron sulfide is very slow in air condition, even if the air humidity is high. The oxidation characteristic of FeS is related to its granularity. The smaller the granularity, the more active the characteristics. When the granularity of FeS increases, the specific area will decrease; thus the contact area between the oxygen molecular and FeS will decrease, which decreases the reaction probability of FeS, so that the oxygen consumption rate of FeS diminishes.

1.3.1.3 Preparations of FeS

Sulfurizing method is commonly used for preparing FeS coating. In the 1970s, France successfully developed the technique of low temperature electrolysis sulfuration, and then this method became the main sulfuration technique at that time. But this method has many environmental protection problems, such as salt bath ageing, absorption deliquescence, decomposition volatilization and stench producing etc. In the 1980s, according to the principle of ion nitriding, Chinese scientists invented a new sulfuration technique-low temperature ion sulfuration, and it has been extensively utilized. Sulfurizing is divided into single sulfurizing and multi-elements co-cementation, such as S—N, S—N—C, and S—N—O. In recent years, Tsinghua University has employed the thermal spraying and sol-gel methods to prepare FeS coatings, and has obtained good results.

1. Sulfurizing

In the 1920s, it was found that the addition of sulfur into lubricating oil could significantly improve the load-carrying capacity of oil and prevent scuffing and seizure of friction-surface. This was due to the formation of FeS film on the friction surface, which can avoid the adhesion and cold-welding of contact points. When the oil with the sulfur-containing additive is used on the surface of iron and steel, the formed sulfide boundary film is extremely thin; it will be worn away rapidly, when the active oil converts to neutral oil. In order to obtain a thicker sulfide film on the friction surface, the sulfuration technology was put forward in the 1940s. The pretreated parts were placed in the caustic soda aqueous solution containing sulfur, and then treated for several minutes at temperatures of 130—140°C, the sulfide film with thickness of several microns was obtained. This film could greatly decrease the scuffing of automobile engine cylinder liners and piston rings. In the 1950s, cyanide salt bath sulfuration (Sulfinuz) appeared in France, the machine parts were treated in the molten salt for 1—3h at temperatures of 560—580°C, the thickness of sulfide layer could reach 0.3mm, but the batch formula of salt bath is private. In the 1960s, the cyanosulphurization method was used to treat parts by more than twenty countries, such as France, Britain, Germany, America, Italy and Japan. The maximum part treated by this method reached three tons, and the effect was good. When the automobile engine piston ring was treated by this method, the wear decreased by 94%; the wear of cylinder liner decreased by 80%. Thus the bronze pieces of some machines even could be replaced by cyanosulfurized iron and steel pieces. In order to meet the ever-increasing requirements of environmental protection and to overcome the shortcoming of severe pollution, the harmless cyanosulfurization method (French Sursulf, American Melonite and Chinese LT, etc.) appeared in the 1970s. Many enterprises established cyanosulfurization production line controlled by computer, e.g., Parkin Company employed this production line to produce 225000 crankshaft of 1045 steel each year. However, the treatment temperature of cyanosulfurization was high; the loss of substrate strength was large. Moreover this method may lead to large deformation of workpieces. To this end, low temperature electrolysis sulfuration (Sulf-BT) appeared. The treatment temperature of the method was around 200°C; the treated pieces were not deformed and soften; almost all iron and steel materials except high chrome steel with the chrome content over 13% can be treated by this method, especially high carbon tool steel, case-hardened steel, nitrided steel, bearing steel, and all kinds of low alloy tool steel. In the 1970s, with the development of environmental protection technology and improvement of electrolyte ageing problem, the low temperature electrolysis sulfuration technology has been industrialized and realized microcomputer controlled automatic production.

Since the 1950s, China began to research the sulfuration technology and imitate the cyanide-free method used by the former Soviet Union, namely, solid packing method and gas method. In the 1960s, China researched the processes of sulfuration, sulfonitriding, cyanosulfurization and low temperature electrolysis sulfuration. A certain accomplishment had been made in the aspects of prolonging service life of cutlery and improving the wear-resistance of tools and parts. However, due to the high investment on the equipment of pollution treatment, the method could not be industrialized. In the 1980s, with the development and maturity of vacuum and plasma techniques, low temperature ion sulfuration appeared. Many units such as Metal and Chemistry Institute of Academy of Railway Science, Shandong University of Technology, South China University of Technology and Machinery Research Institute of Qingdao successively obtained more than 10μm thick sulfide layers by this method, but the industrialization could not be achieved. However, the new concept had been exploited for ion sulfuration. This new technique overcame many shortcomings of traditional electrolysis process, such as low sulfurizing speed, large labor strength, long process cycle, deformation of workpiece and environment pollution caused by the mass usage of toxic mediums; the method had also solved the problems of ageing of electrolysis sulfuration solution and deliquescence of salt bath; therefore, it demonstrated good industrialization prospect.

The sulfuration technology includes the following several kinds of methods according to the processing medium state: solid packing method, salt bath method, gas sulfuration method, electrolytic sulfuration method and ion sulfuration method. As the solid packing method and salt bath method have large labor intensity, high energy-consuming, long period and big workpiece deformation, they were not used any more. The gas method and electrolytic method were toxic and explosive; the processes were unstable; it was difficult to obtain sulfide layers with stable properties; moreover, the industrialized production was difficult to be realized. The low temperature electrolytic sulfuration and low temperature ion sulfuration methods have solved the problem of workpiece deformation. It is possible to achieve industrial application. According to different sulfurizing temperatures, the sulfuration can be divided into low temperature (160—200°C), medium temperature (520—600°C) sulfuration and high temperature sulfuration. Due to the higher tempering temperature, the high-speed steel, high-chromium steel, as well as quenched and tempered steel, etc. can be sulfurized by medium temperature sulfuration, while the carbon tool steel, case-hardened steel, low-alloy tool steel and bearing steel should be sulfurized by low temperature sulfuration. The salt bath of medium temperature liquid sulfuration contains large quantity of highly toxic cyanide, which are harmful to workers and pollute the environment; meanwhile, the cost is high. These unfavorable factors limit its application. For low temperature sulfuration, the evaporation capacity of cyanide salt is small; it is less harmful to people as long as the temperature is well controlled; the application of this method is extensive. All kinds of sulfuration methods are shown in Table 1.2.

Sulfurizing can effectively improve the anti-scuffing and wear-resistance properties of parts, but it can fully play this role only when the part surface possesses high hardness, strength and toughness to give a strong support to the sulfide layer. Therefore, surface quenching and surface strengthening should be combined with sulfurizing treatment. In order to improve the effect of sulfurizing, the following methods are always employed.

Table 1.2 All kinds of sulfuration methods

Classification	Method	Sulfurizing medium	Treated temperature	remark
Solid method	Solid method	Charcoal absorbing H_2S	200—600°C	1—3h
Gas method	Gas method	H_2S	250—600°C	3h
	Gas method	H_2S+H_2	200—600°C	Several hours
Liquid method	Aqueous solution	$(NH_2)_2CS$,	90—180°C	1h
	Salt bath method	NaCN, NaCNO, Na_2SO_3	500—600°C	2—3h
	Electrolytic method	NaSCN, KSCN, etc.	190±10°C	2.5A/dm^2
Ion sulfuration	Ion sulfuration	H_2S, CS_2, S vapor, etc.	160—560°C	1.33Pa, 2h

(1) Chemical heat treatment is first conducted to strengthen substrate surface, and then the surface is treated by sulfurizing, e.g., workpieces are first treated by nitriding, soft nitriding, and carbonitriding and then sulfurized. This method is mostly used for high-chromium stainless steel and heat-resistant steel.

(2) Many composite treatments are commonly used to combine with sulfurizing, such as sulfurnitriding, sulfurnitriding+steam processing, sulfurcarbonitriding, secondary hardening+sulfurcarbonitriding, carburizing+electrolytic sulfurizing, carbonitriding+low temperature sulfurizing, carbonitriding+magnetization electrolytic sulphurizing, O—S—N co-cementation, V—N—S co-cementation, S—Mo co-cementation and S—N—C—O co-cementation (LT process), etc.

Low temperature electrolysis sulfuration, sulfurnitriding and low temperature ion sulfuration are mainly introduced as follows.

(1) Low temperature electrolysis sulfuration.

The temperature of low temperature electrolysis sulfuration ranging from 185°C to 195°C, is lower than heat treatment temperature of parts. Therefore, the change of mechanical properties and deformation are small. The principle diagram of the equipment is shown in Figure 1.6. The basic component of bath salt is the mixed salt composed of 75% KSCN and 25% NaSCN. Its melting point is 125°C. At the temperature of 180—190°C, the mixed salt is entirely ionized

$$KSCN \rightarrow K^+ + SCN^-$$

$$NaSCN \rightarrow Na^+ + SCN^-$$

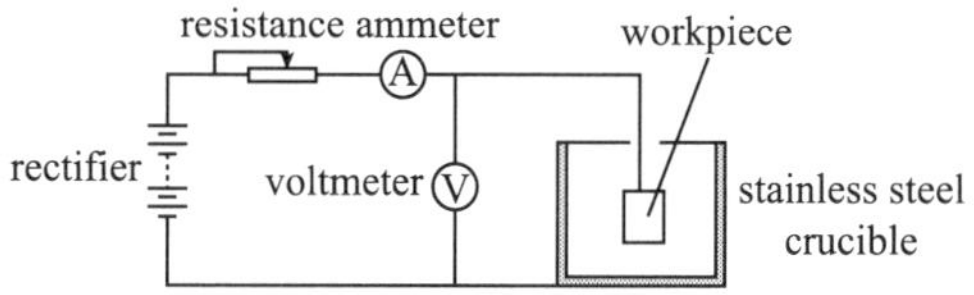

Fig.1.6 Principle diagram of low temperature electrolysis sulfuration

The following reaction occurs during the electrolysis process.
Near the cathode:

$$SCN^- + 2e \rightarrow S^{2-} + CN^-$$

$$K^+ + e \rightarrow K$$

$$Na^+ + e \rightarrow Na$$

Near the anode:

$$Fe \rightarrow Fe^{2+} + 2e$$

$$2SCN^- - 2e \rightarrow 2S + (CN)_2$$

Therefore:

$$Fe^{2+} + S^{2-} \rightarrow FeS$$

The current intensity is 2.5—3.2A/dm^2; electrolysis voltage is less than 4V; after electrifying for 10—20min, the parts are taken out and air-cooled to room temperature; then the parts are cleaned in the oil under the temperature of 80—100°C.

The biggest shortcoming of low temperature electrolysis sulfuration is that the salt bath is unstable and easy to be aged, which will affect the mass production.

Low temperature electrolysis sulfuration has extensively been applied to cutting tools, dies cylinder liners, piston rings, axes, and bearings, etc. FeS layer can not be generated for non-ferrous metals. It should be noticed that the sulfide layer will lose its role if the counterpart of sulfurized surface is copper or copper alloy.

Four-ball, Timken, and Falax testers are usually employed to examine the tribological properties of surfaces after low temperature electrolysis sulfuration under five commonly used extreme pressure lubrication conditions. The results showed that sulfide layer could effectively improve the anti-scuffing property of the surface. However, the experimental results were related with the experimental methods, such as, kind of additive, shape of contact surface and contact pressure. When the experimental method was improper, the sulfide layer could not exhibit excellent properties.

After the hydraulic pressure pieces are treated by low temperature electrolysis sulfuration, FeS layer (HV 50—70) with the thickness of 5—7μm can be formed on their surface to play the role of seizure resistance and anti-friction; especially under oil lubrication condition, the oil film is not easy to be destroyed, and its compression strength can be improved, which are beneficial to reducing the wear, increasing the lubrication, decreasing the noise and improving the seal, so that the service life of hydraulic pressure pieces can be prolonged [2].

(2) Sulfurnitriding.

Sulfurnitriding is a chemical heat-treatment method, which can make sulfur and nitrogen diffuse into workpiece surface simultaneously. The co-cementation process is conducted at 500—600°C. The inner layer is composed of nitrogen compounds and α-Fe, while the external layer is mainly sulfide layer. Sulfide layer has low friction coefficient and strong anti-scuffing properties, while nitrided layer has high strength and good toughness; therefore the co-cemented layer possesses the advantages of sulfide layer and nitrided layer. After W18Cr4V steel is treated by several different heat-treating methods, the depth of diffusion layer and surface hardness are shown in Table 1.3. The study on the sulfurnitrided XC10 steel (0.06%—0.14%C, 0.3%—0.5%Mn, 0.1%—0.4%Si) showed that the surface was composed of ε phase, Fe$_3$N and FeS after the workpiece surface was treated by sulfurnitriding; its properties of anti-scuffing, wear-resistance, fatigue wear resistance and corrosion resistance are all improved.

Table 1.3 Depth of diffusion layer and surface hardness after W18Cr4V steel is treated by several different heat-treating methods

Heat treat method	Depth of diffusion layer/μm	Surface hardness
Conventional heat treat	—	HRC 63—65
Ion sulfuration	110—130	$HV_{0.1}1302$
Low temperature liquid sulfuration	3—5	Before sulfuration: HRC 61—63 After sulfuration: undeterminate
Ion sulfonitriding	110—120 Sulfurized layer 3—5	$HV_{0.1}1280$

The 1045 steel after quenching and low temperature tempering was nitrided at 560°C, and then treated by low temperature ion sulfuration around 200°C. It was found that sulfurizing and nitriding could improve the anti-scuffing performance of surface only at low speed condition, while the composite treatment could significantly improve the anti-scuffing property at various speeds. The wear-resistance of surface by composite treatment was also obviously superior to that by sulfurizing and nitriding; however, the wear of counterface was even more severe than that of original surface.

The gray cast iron was treated by nitriding and sulfurizing successively, and the original surface, nitrided surface and nitrided + sulfurized surface were also examined. The results showed that the gray cast iron after composite treatment possessed concurrently the high hardness of sub-surface, and the friction-reducing property of surface layer, therefore, the tribological properties of the treated surface was obviously better than those of the original material and nitrided layer [3].

(3) Low temperature ion sulfurizing.

The basic principle of ion sulfurizing is similar to that of ion nitriding. The H_2S, CS_2, or vapor of solid sulfur are commonly used as the reacting gases. The workpieces are connected to the cathode; the furnace wall is connected to the anode. When the vacuum degree reaches 1 Torr, high voltage direct current is applied to the cathode and anode. The voltage ranges from 450V to 1500V; the current intensity depends on the surface area and shape of part. When the voltage is more than 500V, sulfur-containing compound gas is ionized to sulfur ions in the electric field of high voltage direct current; then they move towards the cathode and are accelerated near the cathode by the action of cathode pressure drop to bombard the cathode surface; part of the kinetic energy is converted to heat, which will increase the workpiece temperature. Meanwhile, sulfur ions combine with iron atoms/ions sputtered from the workpiece surface to generate sulfide ($Fe^{2+}+S^{2-} \rightarrow FeS$) layer on the workpiece surface. During the process of glow discharge, lots of crystal defects are formed on the surface of iron and steel under the action of ion bombardment, their depth can reach 0.05mm. Obviously, the active surface is beneficial to the generation of sulfide. Although the sulfur atom is big and has low solubility in the iron, it can still diffuse into certain depth along the grain boundaries or defects. The sulfurizing temperature varies with the materials ranging from 140°C to 300°C. When the sulfurizing temperature increases, the thickness of sulfurized layer is obviously enlarged, as shown in Table 1.4. Grey-white color glow is generated when the sulfur-containing gas is ionized, which is the basic characteristic of the normal sulfurizing process. The generation of grey-white color glow is related to the glow voltage, temperature, and gas flow. Relatively ideal glow can be obtained when the voltage is more than 500V.

Table 1.4   Thicknesses of sulfurized layer at different temperature

Sulfurizing temperature/°C	180	200	220	260	300
Thickness of sulfurized layer/μm	6.4	9.6	11.2	19.2	21.0

The substrate is Cr12 steel; the flux of sulfur-containing gas is 150 l/h; the holding time is 1h.

Low temperature ion sulfurizing technique overcomes many shortcomings of low temperature electrolysis sulfurizing, such as ageing of salt bath, absorption deliquescence, decomposition and volatilization, stench producing. Compared with other sulfurizing methods, the advantages of low temperature ion sulfurizing are prominent: ① the sulfurizing temperature is low (200°C), the original hardness, shape and precision of workpieces can be maintained; ② the vacuum treatment does not pollute the surface of workpieces, which do not need to be cleaned after sulfurizing; ③ it is helpful to protect the environment; ④ the thickness and composition of sulfurized layer can be well controlled through adjusting the parameters of voltage, pressure, current and temperature, etc. In recent years, due to its significant advantages, low temperature ion sulfurizing has extensively been applied in various industrial departments, such as metallurgy, automobile, textile, and machine building, etc.

For instance, sulfurized layers were prepared on the 1045 steel with different pre-treatment by low temperature ion sulfurizing [4]. The friction and wear tests for the sulfurized and unsulfurized specimens were compared under dry condition. The results showed that the sulfurized specimens possessed obviously better friction-reducing and wear-resistance properties. This technique has successfully been applied to four kinds of wear-susceptible parts of bottle-making machines.

The wear-resistance of Cr12, GCr15, T10, 1045, and 20CrMnTi steels was improved by 2—3 times when they were treated by low temperature ion sulfuration; meanwhile, their seizure resistance was also greatly enhanced. Figure 1.7 shows the variation of friction moment for the Cr12 steel. The wear life of sulfurized specimens before seizure was significantly prolonged.

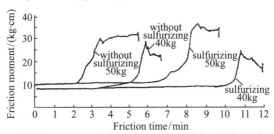

Fig.1.7   Variation of friction moment before the seizure of the Cr12 steel

2. Thermal spraying methods

Thermal spraying technology has the following characteristics:

(1) The process is flexible, and the scope of application is broad. The spraying objects can be small from internal hole with a diameter of 10mm to big bridge and tower. The spraying process can be conducted in the laboratory or plant, in vacuum or atmosphere; it can also be operated in the field.

(2) The substrates and spraying materials are extensive. The substrates can be metals and non-metals, including ceram, plastic, gypsum, cement, lumber and even paper sheet. Spraying materials can be metals and their alloys, plastic and ceramic, etc.; Almost any kind of materials can be sprayed.

(3) The workpieces processed by thermal spraying, except flame spraying and plasma arc powder surfacing, are less heated and have little stress and deformation.

(4) The productivity is high. It ranges from several kilogram (the weight of materials sprayed per hour) to several decades kilogram.

For thermal spraying FeS coating, the gas, liquid fuel or arc, plasma arc and laser as heat source can be adopted to heat FeS particles to melting or semi-molten state; then they are atomized by high speed current, sprayed and deposited on the workpiece surface to form the FeS coating.

Sprayed coatings with different thickness were prepared by plasma spraying method. The FeS particles coated by nickel with the size of 40μm were used as sprayed material. Three coatings with thickness of 0.3mm, 0.8mm and 1.5mm were prepared on the 1045 steel with hardness of HRC 55. The tribological properties of FeS sprayed coatings were examined on a QP-100 friction and wear tester. The result showed that the friction coefficient ranged from 0.11 to 0.14 at a load of 50N; the friction-reduction, wear-resistance and anti-scuffing properties of sprayed coatings were obviously superior to those of the substrate.

3. Sol-gel methods

Since the 1980s, sol-gel technique has been paid increasingly more attentions. Sol-gel is a wet chemical synthesis method and has many characteristics: the equipment is simple; the process is easily controlled; the purity and evenness of products are high; especially the evenness of multi-component products can achieve molecule or atom scale. Therefore, this method is extensively applied to prepare the coatings with special properties of mechanics, chemistry, corrosion resistance and wear-resistance, etc. and functional coatings with properties of light and electrocatalysis, etc.

The basic process of sol-gel technique is as follows: the metal compound, which is easy to hydrolyze, reacts with water in a certain solvent; the gelation is gradually formed through the processes of hydrolysis and polycondensation; the required coating is finally obtained by post treatments of drying and sintering. Alkoxy metals are usually used as sol material. The acid is added to the ethanol solution of alkoxy metal and then decomposed by the addition of water under the action of catalyst; the polycondensation reaction occurs to generate colloid particles; the sol is formed when the polymerization degree of colloid particles decreases.

In the preparation process of sol-gel coating, the sol obtained after the addition of water is the original solution of coating. The viscidity and other rheological behaviours will affect the sol-gel process, and the thickness, evenness and bonding strength of coating. The control of watering amount, sol concentration, reaction temperature and the selection of catalyst and complexing agent are the main factors to impact sol properties. There are lots of methods to prepare sol-gel coating; dip-coating method is commonly used. This method is simple and efficient for the substrate with big area and double-face coating. The sol-gel coating should be heat-treated subsequently: the solution is firstly volatilized by polycondensation; then the sintering treatment is performed to obtain dry coating, which combines well with substrate.

The process of sol-gel method can be shown by Figure 1.8: The solution composed of metal alkoxide, solvent, water and catalyst forms homogeneous sol by hydrolytic polycondensation, and the sol is further aged to form wet gel ①; it is vaporized to obtain dry gel ③, or the solvent is removed after the evaporation to obtain aerogel ②; the dry gel is sintered to obtain dense ceramic body ④; meanwhile, the homogeneous sol can be coated on different substrates ⑤ to obtain even and dense film after roast ⑥; it can also be wiredrawn to obtain glass fiber ⑦ and treated by different method to obtain powders ⑧.

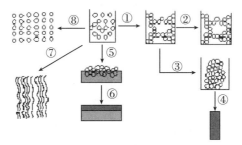

Fig.1.8   The process of sol-gel method

The preparation of sol-gel coating is to mix the base material and FeS powders and stir them evenly; then the special coating equipment is used to coat the sol evenly on the substrate surface; FeS coating is finally formed by drying and curing at room temperature.

FeS coatings with different thickness were prepared by sol-gel method [5]. FeS particles with grain size of 20μm were used in the experiment. The solvent was a patented inorganic silicon water-based paint (ZM-1); it accounted for 10% of the whole sol mass. The filler materials (solute) were the mixture of FeS powders and zinc powders; they accounted for 90% of the whole sol mass. Special equipment was utilized to spray the sol homogeneously onto the surface of 1045 steel about 200μm thick. The result showed that FeS coating could obviously improve the tribological properties of the 1045 steel under oil lubrication condition.

1.3.1.4   Solid lubrication mechanism of FeS

FeS can obviously improve the anti-scuffing, friction-reduction and wear-resistance properties of iron and steel. Its mechanism can be summarized as follows:

(1) FeS has a hexagonal structure and can easily slip along the close-packed planes. Its melting point is 1100°C, thus it can play the role in a wide temperature range.

(2) When the friction pair is in the relative motion, FeS is crushed to adhere to the friction surface and fill into the valleys of worn surface, which can effectively prevent the direct contact between metals and mitigate adhesion.

(3) Sulfurized layer can play the role of "peak load shifting" for the surface to make it smooth and lead to the increase of actual contact area and decrease of contact pressure; meanwhile, the asperities on the metal surface are converted to soft sulfide

to eliminate the source of interior abrasive particles. Therefore, the running-in period is greatly shortened; the friction-pair enters the stable wear stage in advance, and this stage is significantly prolonged to postpone the occurrence of scuffing. In addition, the metal debris may be embedded into the thicker sulfurized layer. Therefore, both the abrasive wear and fatigue wear can be weakened.

(4) Sulfide layer is a porous structure, which is beneficial to easily adsorbing the lubricating oil to form oil film. In addition, FeS can be electrostatically polarized under the action of compressive stress, which can increase the bonding strength between the polar molecule of surface activated substance in the oil and metal surface, and significantly improve the strength of boundary lubricating layer. Therefore, minute quantity of oil can significantly improve the anti-scuffing and wear-resistance properties of sulfurized layer, especially sulfocarbonitrided layer.

(5) During friction, the complicated physico-chemical process will happen on the friction surface. Sulfide is heated in air to generate oxide and precipitate active sulfur atoms. Part of them is oxidized; part of them reacts with metal to form FeS again at high contact temperature, i.e. sulfur can continuously diffuse or migrate into the metal. This is helpful to keep the good characteristics of sulfurized surface for a long time. Therefore, although the thickness of sulfurized layer is only about 10μm, in practice, its role can exceed this range during friction.

(6) Studies have shown that the sulfur element can diffuse into substrate along grain boundaries and crystal defects, it is different from MoS_2 coating; which is formed without metallurgical bonding between MoS_2 coating and substrate. This characteristic of sulfurized layer makes it combine well with substrate and sustain longer time of lubrication efficiency in the process of friction and wear.

(7) High vacancy concentration in the sulfide is beneficial to the diffusion of oxygen and formation of oxide film to avoid scuffing and adhesion.

1.3.2 MoS_2

MoS_2 is the most extensively used solid lubricant at present. There are lots of using methods, such as powder agent, oil solution, aqueous solution and composite material composed of MoS_2 and other metals or high molecular materials. Various solvents and agglomerants can be used to smear or spray the suspensions of MoS_2 on the friction surface to form dry film; the plasma spraying and sputtering methods can also be used to deposit MoS_2 on the friction surface to form thin film.

1.3.2.1   Crystal structure of MoS_2

MoS_2 has a hexagonal layered structure, as shown in Figure 1.9. MoS_2 crystal is constituted by unit layers composed of three planes of S—Mo—S. In the interior of unit layer, each Mo atom is enclosed by sulfur atoms with triangular-prism-shaped distribution; they are connected by strong covalent bond. The thickness of each unit layer is 0.625nm; there are about 40000 unit layers in the thin layer with thickness of 0.025nm. The layers are connected by weak molecular force and MoS_2 is easily to slip between layers; therefore, MoS_2 has excellent solid lubricating properties.

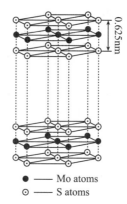

0.625nm

● —— Mo atoms
⊙ —— S atoms

Fig.1.9   Schematic diagram of crystal structure of MoS_2

1.3.2.2   Properties of MoS_2

1. Basic properties

The basic properties of MoS_2 are shown in Table 1.5.

Table 1.5   Basic properties of MoS_2

Performance	Value
Relative atomic mass	160.08
Crystal type	Hexagonal crystal system
Density/($g \cdot cm^{-3}$)	4.5—4.8
Melting point/°C	1185
Adhesivity	Strong bonding force, can not damage the metal surface
Friction coefficient	0.03—0.09 (in air, below 400°C)
Specific resistance/($\Omega \cdot cm$)	851
Heat conductivity/($W \cdot m^{-1} \cdot K^{-1}$)	0.13 (40°C); 0.19 (430°C)
Coefficient of heat expansion/(1/°C)	10.7×10^{-6} (base plane)
Surface energy/($J \cdot m^{-2}$)	24×10^{-3} (base plane) 0.7 (prism plane)
Microhardness (HV)/Pa	3136×10^5 (base plane) 882×10^7 (prism plane)

MoS_2 powder is similar to graphite exhibiting also flat shape. The prism plane of MoS_2 powder has higher activity and is very hard; it can grind metals. The physical properties vary with its crystal orientation: the base plane and prism plane have different surface energy and hardness; the hardness of the latter is 30 times as high as that of the former. The crystal anisotropy is helpful to improve the load-carrying capacity of MoS_2.

2. Thermal stability

The thermal stability of MoS_2 is relatively good. It is an important requirement to lubricants. MoS_2 begins to be oxidized gradually around 400°C in air; after 540°C, the oxidation speed sharply increases to form MoO_3 with large friction coefficient; but during the period when MoS_2 does not become MoO_3 completely, it still possesses lubricating property until 525°C. MoS_2 begins to decompose until 982—1093°C in vacuum, while it begins to decompose until 1350—1472°C in argon. MoS_2 still possesses good lubricating property at low temperature of −60°C. However, the thermal stability of common oil is poor. It becomes thin at high temperature and congealed at low temperatures, which impacts the normal operation of mechanical equipment. In addition, the silicone oil used for aviation can resist the temperature of 250°C; the oil of freezer can resist the temperature of −45°C; their applicable temperature ranges are far lower than that of MoS_2.

3. Chemical stability

The corrosion resistance of MoS_2 is very good; general acids except hydrogen nitrate, aqua regia, boiling hydrochloric acid and concentrated sulfuric acid, do not have effect on it. MoS_2 can be oxidized slowly when the pH value of alkali solution is greater than 10. It is unstable under various strong oxidants and can be oxidized to molybdenic acid, which corrodes the metal surfaces. MoS_2 does not dissolve in the cooling and boiling water. It can keep high chemical stability in the grease, ethanol and ether.

4. Compression property

MoS_2 has layered structure; sulfur atoms and molybdenum atoms combine well. Therefore, its compression property is incomparable with other lubrication materials. The sulfur in the MoS_2 is active; strong sorption can happen between the sulfur and metal atoms on the clean surface. Thus, MoS_2 has strong bonding strength with the metal surface and can not be easily destroyed during friction, so that it can bear relatively high load. MoS_2 film with thickness of 2.5μm can bear the contact pressure over 2800MPa and friction speed of 40m/s. MoS_2 can also be used as extreme pressure additive, which can be added to grease or suspended aqueous solution. The life of ball bearings lubricated with MoS_2 is more than 1500h at speed of 2000—4000r/min; the decrease of life is small, even if the load is increased from 19.6N to 196N.

5. Adhesive property

As sulfur atoms combine well with metal, strong sorption occurs between MoS_2 and metal; extremely thin MoS_2 can play a good role of lubrication and keep comparatively long life.

6. Effect of humidity

The friction coefficient rises with the increase of humidity in air; when the relative humidity is less than 15%, MoS_2 has the best lubricating effect. Therefore, the relative humidity should be less than 15% when MoS_2 is used.

7. Resistance to high vacuum

The lubricating oil will vaporize in high vacuum and has defective impact on the substance in vacuum, while MoS_2 is different from the oil. The friction coefficient of MoS_2 in vacuum is even lower than that in air; MoS_2 still has good lubricating property under ultra-high vacuum condition. Therefore, MoS_2 is an effective lubricating material under ultra-high vacuum condition.

8. Electromagnetic property

MoS_2 is poor conductor and non-magnetic material at room temperature.

9. Radiation-resistance property

MoS_2 possesses anti-radiation property; its normal lubrication can not be destroyed during radiation. Thus, it is an indispensable lubricant in the modern national defence industry and frontier sciences and technologies, such as spacecraft, artificial satellite, and atomic reactor, etc.

1.3.2.3 Preparation of MoS$_2$ film

1. Bonding method

Bonded solid lubricating film is one of the main methods of preparing solid lubricating coatings; it has been extensively applied in industries. Bonded solid lubricating film is also well known as dry film lubricant. The powders of solid lubricants (MoS$_2$ or PTFE) are dispersed in the system of organic agglomerants (phenolic aldehyde, epoxy resin, silicone, polyimide, polyphenylene parallel thiazole and polyphenylene parallel imidazole etc.) or inorganic agglomerants (Na$_2$SiO$_3$, K$_2$SiO$_3$, B$_2$O$_3$, Na$_2$B$_4$O$_7$, Na$_3$PO$_4$, and K$_2$PO$_4$, etc.), then the "paint" is smeared or sprayed on the friction surface with thickness of micron scale, after drying, the solid lubrication bonding film is obtained.

The thickness of dry film generally ranges from 20μm to 50μm; the maximum thickness can be greater than 100μm. The friction coefficient of dry film generally ranges from 0.05 to 0.20; the minimum can reach 0.02. Dry film has the same load-carrying capacity as substrate, because the load is applied to the substrate through dry film. If the dry film is too thick, it will be crushed so that the load-carrying capacity and wear-resistance properties will be decreased.

Bonded solid lubricating film can be used under many environment conditions, such as high temperature, heavy load, ultra-low temperature, strong oxidation-reduction and radiation without pollution. Dry films can be got from markets, from extremely low temperature below $-200°$C to high temperature of about $1000°$C. For instance, the service temperature of epikote-system bonded dry film is -70—$250°$C, while that of polyimide-system dry film is -70—$380°$C. In addition, within the service temperature range, the dry films do not have phase transformation; the friction coefficient is also relatively stable. Dry films possess excellent corrosion resistance and dynamic sealing properties; it can also prevent mechanical vibration and noise. However, the dry films do not possess cooling role; therefore, they can not be applied to machine components with high sliding speed.

(1) Organic bonded solid lubricating film.

The commonly used organic resins are alkyd resin, polyurethane, epikote, phenol formaldehyde resin, polyimide resin and its modified products, polyacrylate, organic silicone resin, aromatic heterocyclic polymer and other thermoplastic resins. For the convenient smearing, organic solvent and water are generally used for diluent. In recent years, people have successfully developed the solvent-free and powder sprayed bonded dry films.

① Epikote bonded dry film.

Table 1.6 shows the recipes of HNT coating series with the agglomerant of epikote, solidification agent of propylene oxide-butyl ether, additives of Di-n-butyl phthalate, various additives and solid lubricants. After solidifying for over 24h at room temperature, they can be used immediately. In order to make the coating adhere well with substrate, the pressure about 0.1MPa should be applied on the coating during its solidification. The practical application showed that when the HNT coating was sprayed on the surface of guide way of a planer milling machine, the annual wear loss was only 5—7μm under normal condition of operation.

Table 1.6 Recipes of HNT epoxy coatings

Component \ Recipe	HNT11-15	HNT17-5	HNT20-1	HNT21-4
	Addition/g			
Epikote	100	100	100	100
Di-n-butyl phthalate	10	10	15	15
Propylene oxide-butyl ether	12	10	10	15
Gas phase silicon oxide	2	1	2	1
Ferrous powder	25	15	25	15
Titanium oxide		30	15	30
Molybdenum	100	80	80	80
Graphite	25	20	20	20
Total amount	274	266	267	276

② Polyimide bonded dry film.

Polyimide possesses excellent mechanical properties at temperature ranging from −200°C to 260°C, chemical stability and radiation resistance as well as high bonding strength with metals; therefore, polyimide can be used as agglomerant for high temperature-resistant, radiation-resistant and corrosion-resistant dry films. However, the synthetic process is complicated; the cost is high; it is difficult to machine. Therefore, its application is limited. To this end, the poly-amino-bimaleiraide resin is synthesized. This resin has good characters as polyimide resin; the synthetic process is simple; the material source is extensive, and the price is low. The most important thing is that it can be dissolved in some organic solvents. For instance, the bonded dry film with the agglomerant of poly-amino-bimaleiraide, solid lubricant of MoS_2 and solvent of diethylenelmide possesses good comprehensive properties.

The dry film containing agglomerant of modified polyimide resin and solid lubricant of MoS_2 with addition of Sb_2O_3 is solidified for 2h at high temperature; the obtained dry film is known as PI dry film. Its composition is: polyimide: MoS_2:Sb_2O_3= 1:3:1. This kind of dry film has low evaporation rate; it is compatible with liquid hydrogen and liquid oxygen; it can be used under temperature ranging from −178°C and 300°C and under vacuum condition of $133.322×n$Pa; its wear life is 270m/μm.

③ Polyphenylene sulfide bonded dry film.

Polyphenylene sulfide resin (PPS) is white powder and has linear structure with high crystallinity; it is a kind of new thermoplastic synthetic material. It can be cured and crosslinked within temperature of 320—390°C, and can not be dissolved in any commonly used solvents below 170°C; thus it can be sprayed only using suspended agent. Ethanol is a good dispersant for PPS, but absolute alcohol is easy to volatilize when it is used; therefore, certain proportional water should be added. Experience shows that the optimum volume ratio of water and ethanol is 4:10; due to the decrease of ethanol volatilization amount in winter, the ratio can be changed to 4:8. The PPS bonded dry film was sprayed on the substrate of stainless steel and solidified for 1h at temperature of 370°C; then it was cooled by a furnace cooling mode. The obtained bonded film is known as PPS-1 dry film. The tribological properties were examined on a TimKen tester at the load of 315N and sliding speed of 2.5m/s. The measured friction coefficient ranged from 0.08 to 0.11. The wear life was prolonged with the increase of film thickness at a quite wide range of film thickness (70—130μm), while the average wear life was 338m/μm.

(2) Inorganic bonded solid lubricating film.

The agglomerants of inorganic bonded solid lubricating films are inorganic salts such as silicate, phosphate, borate, ceram and metal. Compared with organic bonded film, the prominent advantages of inorganic bonded film are wide service temperature, radiation-resistance, and good compatibility with liquid oxygen and liquid hydrogen. But, the brittleness is large; the load bearing capacity is poor; the tribological properties are inferior to those of organic bonded film. Therefore, most inorganic bonded films are only used in special conditions. For instance, it can be applied in space equipment using liquid oxygen and liquid hydrogen mediums, at particularly high temperature and free of organic vapor pollution conditions.

① Silicate bonded dry film.

The bonded dry film with the agglomerant of potassium silicate, solid lubricants of MoS_2 and graphite, diluent of water is known as SS-2 dry film. The composition is: potassium silicate (anhydrous): solid lubricants=1:4 (mass ratio), MoS_2:graphite=4:1 (mass ratio). A small quantity of distilled water is added to potassium silicate to dissolve it entirely. When the solid lubricants are added and mixed to paste, the addition of water can be 50% (mass fraction) of solid lubricants. During spraying, the addition of water can be 96% (mass fraction) of solid lubricants. The optimum thickness of sprayed film generally ranges from 30μm to 50μm; it is dried at 150°C.

SS-2 dry film is applicable to work at temperature ranging from −178°C to 400°C. 40—50μm thick SS-2 dry film was sprayed on the stainless steel. The tribological properties were tested on a TimKen tester at a load of 315N and speed of 2.5m/s. The measured friction coefficient was 0.06—0.08; the average wear life was 120m/μm.

The bonded dry film with the agglomerant of potassium silicate, solid lubricants of MoS_2, graphite and Ag powder, and diluent of water is known as SS-3 dry film. Its preparation process is similar to that of SS-2 dry film; the drying temperature is also 150°C. The friction coefficient of SS-3 dry film is 0.06—0.08; the average wear life is 176m/μm. It is obvious that the wear-resistance of SS-3 dry film is superior to that of SS-2 dry film.

② Phosphate bonded dry film.

Phosphate bonded dry film was developed for the temperature ranging from room temperature to 700°C. Phosphate is used as agglomerant; graphite, graphite fluoride and BN are used as solid lubricants; water is used as diluent. The tribological properties of phosphate bonded dry film were examined on a high temperature tester at the load of 50N and rotation speed of 1000r/min (1.31m/s). The friction coefficient was very low at high temperature (650—700°C), but the wear-resistance was very poor. The film thickness had little effect on the tribological properties, because the dry film had been softened at high temperature; except a thin film formed on the friction surface, the rest had been squeezed out from the wear track.

In order to improve the wear-resistance of dry film, the spraying agent need a surface activation pretreatment before the dry film was sprayed, otherwise, if the untreated spraying agent was used, the lubricant particles would be easy to form large agglomeration. After the activation pretreatment, the structure of dry film was homogeneous and dense, its strength and wear-resistance were improved.

The fretting tribological characteristics of MoS_2 bonded solid lubrication dry film were investigated through the radial and tangential fretting tester [6]. The results showed that the MoS_2 bonded dry film had good radial fretting wear resistance; there was no mixed zone existed in the tangential fretting condition; the damage of dry film was strongly dependent on displacement amplitude.

2. Vapor deposition method

The MoS_2 solid lubricating ultra-thin films with the thickness ranging from several hundreds nm to several microns can be prepared by vapor deposition techniques, including physical vapor deposition (PVD), chemical vapor deposition (CVD) or plasma chemical vapor deposition (P-CVD). Compared with the sprayed or bonded film, MoS_2 deposited film has many prominent characteristics, such as even composition and thickness, good repeatability, strong bonding strength between film and substrate, low friction coefficient and long service life. These characteristics can meet the special requirements of aerospace, electronic, and precious machinery industries.

(1) Sputtering method.

Sputtering is a process of using energetic ions (positive ions) to bombard the target surface, sputter out the atoms/ions and deposit the film with the same composition as the target on the substrate. Sputtering can be classified into magnetic sputtering and ion beam sputtering.

① Magnetic sputtering method.

Magnetic sputtering is a kind of "high speed and low temperature sputtering technique", which was rapidly developed in the 1970s; it has extensively been applied to the deposition of MoS_2. In vacuum condition, when the electrons are accelerated to fly to substrate under the action of electrical field, they collide with argon atoms to ionize lots of argon ions and electrons, which fly to the substrate. Argon ions are accelerated to bombard the target under the action of electrical field to sputter out a mass of target atoms; the neutral atoms (molecules) are deposited on the substrate to form film. When the secondary electrons are accelerated to fly to the substrate, they are influenced by Loren magnetic force and restrained near the target surface. As the plasma density is high in this zone, the secondary electrons encircle the target surface to make circular motion under the action of magnetic field. The movement route of electrons is long, and they continually collide with argon atoms to ionize lots of argon ions to bombard the target; after multiple collisions, the energy of electrons gradually decreases so that they get rid of the bondage of magnetic line, keep away from the target and deposit on the substrate finally. The characteristics of magnetic sputtering are to change the motion direction of electrons, improve the ionization rate of working gas and effectively make use of the energy of electrons and extension of movement route of electrons.

The tribological properties of MoS_2 film prepared by magnetic sputtering were investigated under four different conditions [7]. MoS_2 with the purity of 99.99% was used as target; and a small quantity of rare metal was also contained. The results showed that the friction coefficient of the specimen with MoS_2 film decreased by three to four times, compared with that of the specimen without MoS_2 film, the friction coefficient of specimens in vacuum was lower than those in air.

It is difficult to use pure MoS_2 film for a long time in air, especially in humid air; MoS_2 is easily oxidized to MoO_3 to decrease its service life. In recent years, the preparation of composite film containing MoS_2 is increasingly paid attentions. The tribological properties of composite film are obviously superior to those of single MoS_2 film.

MoS_2/Ti nano composite film deposited on the SKD-11 steel surface by magnetic sputtering can greatly decrease its friction coefficient and improve its wear-resistance [8]. To increase the film thickness properly is helpful to improve the wear-resistance of Steel.

MoS_2-Cr film prepared on the high speed steel had relatively low friction coefficient and long service life; meanwhile, the wear scar width and depth on the counterpart were low as well [9].

② Ion beam sputtering method.

Under low pressure, the argon ions emitted from ion source bombard the target in a certain angle. As the energy of bombarding ions is about 1keV, the penetration depth on the target can be neglected; the cascade collision can occur only in the surface layer with deepness of several atoms; a mass of atoms escape from the target surface to become sputtering particles with the energy of 10eV order of magnitude. As there is a small amount of background gas molecules in the vacuum chamber, the free paths of sputtering particles are very large, these particles reach the substrate by the line trajectory and deposit on it to form film. During the film formation process, especially the sputtering particles with the energy of more than 10eV can infiltrate the film to several atoms magnitude, so the adhesive strength of the film can be improved. Compared with vacuum evaporation, sputtering deposition has the following characteristics [10]:

a) Sputtering deposition depends on the momentum exchange to make the atoms and molecules of solid material convert to gas; the average energy of sputtering is about 10eV, which is more than 100 times as high as that of vacuum evaporating particles; after the deposition on the substrate, there is still enough kinetic energy for the film to transfer on the substrate surface. Therefore, the film has good quality and combines well with the substrate.

b) Any materials, even the high melting materials can be sputtered, so the sputtering method is extensively applied.

c) The incident ions for sputtering are generally obtained by gas discharge method; the working pressure ranges from 10^{-2}Pa to 10Pa. Thus, the sputtering ions has collided with the gas molecules in the vacuum chamber before the ions reach the substrate surface, their direction of motion randomly deviates the original direction; meanwhile, the sputtering is generally produced from larger target area. Therefore, comparing with vacuum deposition, the film with more homogeneous thickness can be obtained, but the sputtering can make the film contain more gas molecules under high pressure.

d) The sputtering deposition, except magnetic sputtering, has low deposition rate; the equipment is more complicated than that of vacuum evaporation; the cost is relatively high, but the manipulation is simple; the process repeatability is good; the automatic control of process is easily realized. Sputtering deposition is more suitable for the continuous production of high and new technology products, such as large-scale integrated circuit disk, optical disk, and large area and high quality coated glass.

MoS_2-Ti composite film was prepared on the surface of titanium alloy by combined technique of ion beam enhanced deposition and ion beam sputtering deposition; it had higher bonding strength and better density than pure sputtered film; meanwhile, the durability of film was also improved, but the friction coefficient was increased [11].

(2) Pulsed laser deposition method.

Pulsed laser deposition (PLD) was first applied to the preparation of high-temperature superconducting materials. With the progress of researches on the PLD technique and development of high power pulsed laser technique, the scope of films prepared by this method is becoming increasingly extensive. PLD is to focus the high strength pulsed laser beam generated from the excimer pulse laser on the target surface, so that high temperature and ablation can occur on the target surface, high temperature and high pressure plasma ($T \geqslant 104K$) is further generated, and the film is finally deposited on the substrate. The deposited film can be epitaxial layer or amorphous film, which depends on projection rate, substrate temperature and structure. MoS_2/Al composite lubricating film prepared by pulsed laser method can improve the wear-resistance in humid environment [12].

3. Thermal spraying method

Thermal spraying is a surface modification technique; since German invented this method in 1982, it has rapidly been developed for over 25 years. Special equipment is adopted; the heat sources of plasma, arc or flame, etc. are used to heat the wire or powder spraying materials to molten or semi-molten state, which are sprayed by the compressed air from the spraying gun on the substrate surface to form the coating with a certain thickness. $MoS_2/metal$ lubricating film has been prepared by flame spraying.

When the polymer containing MoS_2 is sprayed on the metal surface, the MoS_2 lubricating coating with high adhesion strength and controllable thickness can be obtained for the composite coating prepared by this technique with polyamide as spraying material, and MoS_2 as solid lubricant, when the content of Mo was less than 3%, the composite coating showed high hardness and good wear-resistance [13].

4. Electrodepositing method

In recent years, the electrodepositing technique has greatly been developed, and the deposition methods are becoming increasingly diversified, it mainly includes DC electrodepositing, pulsed electrodepositing, jet electrodepositing and composite electrodepositing, etc.

Electrodepositing method has been adopted to prepare $MoS_2/metal$ lubricating films, i.e., MoS_2 particles are evenly suspended in the plating solution; the particles and metal are co-deposited on the surface of workpiece to form the lubrication film by composite electrodepositing process. This method has many advantages: the preparation can be performed at room temperature; large area deposition is easily realized; the composition and thickness of film can be easily controlled; the manipulation is simple and safe. In addition, electrodepositing can also make the composition of film exhibit a gradient distribution.

5. Particle sandblasting method

The sandblasting of MoS_2 is to spray MoS_2 particles on the workpiece by the high-speed compressive airflow; high purity MoS_2 can be smeared on the sliding surface without agglomeration. The friction-reduction and wear-resistance effects are obvious; the cost is low and the treatment time is short. During the treatment, the emissions are only air and cracked MoS_2 particles, which can be reclaimed. From the point of view that this method does not produce waste totally, and protects the environment; it does not need equipment for waste treating, and reduces equipment cost, this technique is acceptable.

6. Thermal diffusion method

Thermal diffusion method is to heat the material on the metal surface and make it bond with metal surface to form film through increasing temperature to intensify the diffusion. Using this method to prepare MoS_2 film is easily manipulated; meanwhile, the cost is lower than that of sputtering or gas deposition. When MoS_2 solid lubricating film was deposited on the surface of high speed steel and hard alloy, it can prolong the life of hard alloy cutting tool by 1.5 times [14]. In order to improve the bonding strength of MoS_2 film, MoO_3 was heated on the steel plate, and then was sulfurized to form MoS_2 film. The film was thicker than that of commonly deposited film; meanwhile, the bonding strength was higher [15].

1.3.2.4   Preparation methods of nano MoS_2

There are many methods to prepare nano MoS_2, mainly including chemical method, physical method, and single-layer MoS_2 re-stacking method arisen recently [16].

1. Chemical method

According to reaction type, chemical method mainly includes reduction method, decomposition method, oxidation method and electrochemistry method, etc.

(1) Reduction method.

Mo (VI) and Mo (V) can be reduced to Mo (IV) by proper reductants, such as hydrazine type (like N_2H_4), hydroxylamine type (like NH_2OH) and hydrogen; the granularity of prepared MoS_2 can achieve nano-scale through controlling the reaction condition. N_2H_4 can directly reduce $(NH_4)_2MoS_4$ to MoS_2; besides, in the pyridine solution, N_2H_4 can also reduce MoO_3 to MoO_2 by thermal solvent synthesis method, and crystal MoS_2 with grain size of 100nm can be obtained after the addition of sulfur. Hydroxy ammonium salt can reduce the Mo (VI) of MoS_4^{2-} or MoO_4^{2-} to Mo (IV); then the MoS_2 with grain size below 100nm can be prepared. In addition, the carbinol solution of NH_4SCN can reduce MoO_4^{2-} to MoS_2 by the impregnation technique; the obtained film is even nano MoS_2 with hexagonal structure; the band difference of film is 1179eV. By the XRD analysis, it was found that the film prepared below 300°C is amorphous state, while the film prepared under temperature ranging from 360°C to 450°C is crystal state, and the crystallization at 450°C is better than that at 360°C.

The different surface activating agents have a certain effect on the nano MoS_2. The three surface activating agents of anionic dialkyl dithio ammonium salt, cationic cetyl trimethyl ammonium chloride and non-ionic polyvinyl alcohol can decrease the agglomeration degree between particles and give favorable effect on the generation of nano-particles; among them, cationic surface activating agent has the best modification effect, the even MoS_2 particles with the granularity of 100nm can be obtained; meanwhile, they exhibit hydrophobicity and have good dispersibility in the organic phase.

(2) Decomposition method.

Decomposition method is to use thermal decomposition or other high energy physical measure to decompose ammonium thiosulfate or molybdenum trisulfide to MoS_2, which generally needs the protection of inert gas. $(NH_4)_2MoS_4$ and MoS_3 can be prepared by the following ionic reactions:

$$MoO_4^{2-} + 4S^{2-} + 8H^+ = 4H_2O + MoS_4^{2-}$$

$$MoS_4^{2-} + 2H^+ = H_2MoS_4$$

$$H_2MoS_4 = MoS_3 + H_2S \uparrow$$

In the high pure argon airflow, $(NH_4)_2MoS_4$ is thermally decomposed by proper heating rate, and differential thermal analysis and other means are employed to prepare MoS_2, the detailed process is as follows:

$$(NH_4)_2MoS_4 \xrightarrow{160°C} (NH_4)HMoS_4 \xrightarrow{173—180°C} H_2MoS_4 \xrightarrow{180—190°C}$$

$$MoS_3 \xrightarrow{190—220°C} M_2S_5 \xrightarrow{230—260°C} MoS_2$$

Besides the simple thermal decomposition, using the high energy physical means to promote decomposition is also an effective method to prepare nano MoS_2; during the process of chemical preparation, γ ray, ultrasonic wave and other high energy physical means are employed to obtain nano-particles. This method can remedy the shortage of single chemical method, and has high application value. $(NH_4)_2MoS_4$ is used as reactant; the distilled water is used as solvent; polyvinyl or sodium dodecyl sulfate is used as surface activating agent; isopropanol is used as detergent; ammonia is used to adjust the pH value, and high pure N_2 is injected into the solution to eliminate oxygen; then γ ray is adopted to irradiate; finally the grain size of obtained MoS_2 can reach 10nm. The nano MoS_2 is amorphous particles with high stability; the crystallinity is still lower after the heat treatment at 650°C.

(3) Oxidation method.

MoS_2 is prepared by the self-oxidation of $MoCl_3$; the reaction is as follows:

$$4/3MoCl_3 + 2Na_2S \rightarrow MoS_2 + 1/3Mo + 4NaCl$$

The simple sulfur and metal molybdenum can also be directly synthesized to MoS_2. For instance, 0.03μm molybdenum powder and 45μm sulfur powder are used to prepare MoS_2 powder with the grain size below 1.0μm by self-propagating high temperature synthetic method. However, the melting point, boiling point and hardness of metal molybdenum are all high; the reactive activity is rather low. Therefore, the direct combination needs high temperature, and the grain size of product is large. The oxidation of $Mo(CO)_6$ and other low valence complexes overcome the above shortcomings; the particles with small grain size are easily obtained or the thin film is easily deposited. After the metal molybdenum is synthesized to carbonyl complex, the σ—π bond (σ coordinate bond and π feedback bond) is formed to change the electron distribution of molybdenum atoms and make them active to enhance the reactive activity. Meanwhile, the melting point, boiling point and hardness of carbonyl complex can be greatly decreased, and it can easily deviate from the original system containing impurity; the product has high purity, and the reaction can be conducted in the liquid phase or gas phase. For instance, $Mo(CO)_6$ can react with sulfur in the organic solvent to prepare nano MoS_2 with high purity; the grain size ranges from 10nm to 30nm. The shortage of this method is the strong toxicity of molybdenum carbonyl; CO in the molybdenum carbonyl easily combines with hemoglobin in the human body; even the metal can also be taken into human body. This is harmful to the environment and human health.

(4) Electrochemistry method.

The oxidation-reduction reaction in the solution can be achieved through galvanic cell composed of proper device and electrolyte solution to make chemical reaction process on the electrode. Under the action of applied voltage, the plating or electrolyzing reaction may happen on the electrode, which can't be conducted spontaneously originally. The generated sediment through electrode reaction can be deposited on the electrode surface; if proper material is selected as electrode, MoS_2 deposited film can be obtained on the material.

2. Physical method

Physical method is a process of using mechanical milling and high energy physical means to smash, cut or spray MoS_2 to refine or obtain coatings. Natural mineral purification method is to use mineral containing MoS_2 as raw material; then flotation and chemical leaching are employed to remove various impurities in the mineral; consequently, refinement can be achieved through smashing; the obtained product still keeps natural lattice of MoS_2, but the grain size is large. According to milling mode, mechanical grinding can be classified as agitating milling, vibrating milling, ball milling, annular gap milling and air jet milling, etc; among them, in the aspect of air jet milling, China has produced MoS_2 ultra-fine powders with the grain size exceeding the Climax standard (0.55—0.8μm). High-voltage electric arc method can be employed to prepare IF-MoS_2 thin film. In the condition of local high pressure and existence of nitrogen, arc is used to cut solid MoS_2 target to obtain nano MoS_2 film; nano MoS_2 powders with different shapes can be obtained by laser cutting method in the protection of argon.

3. Single-layer MoS_2 re-stacking method

Single-layer molybdenum disulfide re-stacking method (re-stacking method for short) is a special chemical method developed rapidly recently; it is an effective method to prepare two-dimensional nano composite material—MoS_2 intercalation compound (MoS_2-IC). Re-stacking method is generally divided into three steps:

$$MoS_2 \xrightarrow{\text{direct intercalation}} MoS_2\text{-IC} \xrightarrow{\text{separation layer}} \text{single-layer } MoS_2$$
$$\xrightarrow{\text{re-stacking method}} \text{new } MoS_2\text{-IC}$$

In the first step, the more mature method is n-butyl lithium reducing MoS_2 method:

$$xC_4H_9Li + MoS_2 \rightarrow Li_xMoS_2 + x/2C_8H_{18}$$

The generated Li_xMoS_2 can produce the separation layer in the separation reagent:

$$Li_xMoS_2 \xrightarrow{+H_2O} [MoS_2]^{x-} + xLi^+$$

The hydrogen element in the separation reagent is converted to hydrogen gas; the gas expansion leads to the desquamation between layers so as to obtain stable single-layer suspension, actually single-layer $[MoS_2]^{x-}$. Re-stacking can take place for the single-layer MoS_2 in a certain condition; proper object substance can be interfused during re-stacking. Making use of that the object substance is difficult to dissolve the separation reagent, or sediment can be generated through changing condition, we can make the object substance and single-layer MoS_2 co-deposit on the carrier (e.g., aluminum oxide) to obtain various MoS_2 intercalation compound (MoS_2-IC).

1.3.2.5   Preparation method of IF-MoS$_2$

By the elicitation of C_{60} and carbon nanotube, in 1992, Israel Wazman Academy of sciences discovered WS_2 nano-particles and nanotubes with structure of inorganic fullerene; this important discovery uncovered the research prelude of IF nano-material field. In 1993, they synthesized IF-MoS$_2$ nano-particles and nanotubes by the similar method. Subsequently, IF-structured nano-particles and nanotubes of various layered inorganic compounds such as BN, Al_2O_3, $CdCl_2$, $MoTe_2$, and TiO_2 were found one after another. IF-MoS$_2$ nano-particles can be used as solid lubricant and additives for oil and grease in the harsh conditions.

When IF-MoS$_2$ nano-particles were found, their particular microstructure, physical properties and potential application in various fields have caused extensive attentions of the scientists in the world. In order to find more applications, many research groups in the world are devoting themselves to seek effective methods to synthesize IF-MoS$_2$ nano-particles and nanotubes. Now, many methods can be used to synthesize IF-MoS$_2$ nano-particles as follows [17].

1. Arc discharge method

In the de-ionized water, hollow metal Mo stick filled of 2H-MoS$_2$ powders was used as anode to contact the fixed cathode graphite stick; thus the arc discharge (30A) and blue plasma ejecting happened. The temperature was very high (2700°C) during the arc discharge, which led to the consumption of Mo stick. The change of distance between the cathode and anode caused the instability of plasma; in order to keep the stable plasma, anode metal Mo stick should be used continuously to contact the cathode graphite stick to trigger arc. At the beginning of discharge, powdery material was floated on the water surface due to the quenching of plasma in the water. After these powdery materials were collected and dried in vacuum, IF-MoS$_2$ nano-particles were obtained.

2. Pulse crystallization method

At first, $(NH_4)_2Mo_3S_{13} \cdot xH_2O$ was treated by heat decomposition at 400°C to generate amorphous α-MoS$_3$ precursor; then MoS$_2$ nano-particles were obtained after the crystallization of α-MoS$_3$ precursor. One of the crystallization methods was that the precursor was acted for several seconds by electric impulse generated by the pinpoint of scanning tunneling microscope (STM); then the amorphous α-MoS$_3$ was decomposed and crystallized so as to form the core-shell structure with α-MoS$_3$ coated by MoS$_2$ layer; IF-MoS$_2$ nano-particles with cage-like structure was formed finally, and sometimes MoS$_2$ nanotubes could be generated as well. Another crystallization method was that using the high temperature and high pressure generated by ultrasonic electrochemical probe to induce the decomposition and crystallization of α-MoS$_3$ in the solution; IF-MoS$_2$ nano-particles were obtained finally. The researches showed that the processes of decomposition and crystallization were very fast; the whole re-action did not require the environment to supply heat, it could be processed by the energy generated by self-reaction; the shape and crystallization of obtained IF-MoS$_2$ nano-particles were complete.

3. Solution method

China University of Science and Technology had prepared IF-MoS$_2$ nano-particles by hot-solution method. 0.515g Na$_2$MoO$_4$ was dissolved in 10mL de-ionized water and mixed with 30mL hydrazine and 86% water to compose solution A; 1.5mL octanol, 0.5mL CS$_2$ and sodium sulfite dodecane were mixed to solution B; solution A and B were mixed and put in a sealed Teflon pressure cooker, then the mixed solution was heated to 140°C with heating speed of 0.5°C/min, held for 24h and cooled to room temperature naturally; the mixed solution was filtrated; the sediment was washed for several times by ethanol and de-ionized water, and dried for 4h at 60°C in vacuum; black powdery material was finally obtained. Afterward, this material was put in the quartz tube, heated to 800°C and held for 12h in the atmosphere of argon; the product was IF-MoS$_2$ nano-particles.

4. Chemical vapor reaction method

IF nano-particles were first prepared by chemical vapor reaction method, which is a most mature method at present. In the 1993, Israel Wazman Academy of Sciences used the electron beam to bombard the MoO$_3$ film deposited on the quartz substrate, to make the MoO$_3$ film composed of isolated MoO$_3$ nano-particles, i.e., part of the film was discontinuous. The film was heated in the reduced atmosphere (5%H$_2$+95%N$_2$) to make it react with H$_2$S; it was observed by TEM every period of time. The result showed that MoO$_3$ film was gradually sulfurized to MoS$_3$ when the heated film was in the atmosphere of low temperature and excessive sulfur. When the temperature was further heated to 850°C, MoS$_3$ would lose a sulfur atom; finally, MoS$_3$ nano-particles were crystallized to IF-MoS$_2$ nano-particles. Its formation process can be expressed as follows:

$$MoO_3 + xH_2 \rightarrow MoO_{3-x} + xH_2O$$

$$MoO_{3-x} + 3H_2S \rightarrow MoS_3 + (3-x)\,H_2O + xH_2$$

$$MoS_3 \rightarrow MoS_2 + S$$

1.3.2.6   Application of MoS$_2$

Molybdenum chemical industry was on the horizon in the beginning of 19th cen-tury; in the middle of 20th century, with the rapid development of petroleum, auto-mobile, light and mechanical industries, the molybdenum chemical and molybdenum fine chemical industries were promoted to be further developed and extended to deeper field. MoS_2 is an important product in molybdenum chemical industry and has special physical and chemical characteristics; it is entitled as "king of lubrication" in the lubricating field. The application of MoS_2 in the fields, such as hydrodesulfurization catalyst of petroleum, photoelectrochemical cell, non-aqueous lithium battery, high-elastic new material and coating, etc. is the extensively researched subject [18].

1. Usage as bonded solid lubricating film

Bonded solid lubricating film has excellent solid lubricating characteristics; it can make solid lubricant strongly adhered to the substrate surface to prevent the adhesion of metal friction-pair effectively. It can play the role of lubrication in the special environment, where the traditional oil and grease are not suitable to be used. Therefore, it can be extensively applied in vacuum, high and low temperature, inflammable, explosive, strong oxidizing and other environments.

According to literature reports, MoS_2 solid lubricating film has good wear-resistance property; under the same condition, the friction coefficient of MoS_2 bonded solid lubricating film in vacuum is about 1/3 of that in air, while the wear life in vacuum is several times or several tens times as high as that in air. Therefore, MoS_2 bonded solid lubricating film is the first choice of lubrication materials for machinery working in vacuum.

2. Usage for improving the lubricating property of composite material

MoS_2 can improve the rigidity, hardness and lubricating properties of thermoplastic resin and has little effect on the property of thermosetting resins. Because the MoS_2 based self-lubricating composite material containing high-melting metal has high mechanical strength, low friction and wear-resistance characteristics, it is applicable to prepare friction-reducing components under special conditions. The American Boeing Company and Lanzhou Institute of Chemical Physics of China Academy of Sciences have successively conducted researches in this aspect systematically; among them, the latter has successfully prepared $Ag-MoS_2$ system composite lubricating bearing material, which can be used for power transmission and signal transferring devices of outer space. $MoS_2-W-Mo-Nb-Ta$ system self-lubricating composite material used as guide ring material of high temperature section (950K) of refrigerator has successfully been applied; this material can be also used for high vacuum rolling bearing cage and high temperature dynamic seal, etc.

3. Usage as additives for oil and grease

MoS_2 used as oil and grease additives has extensively been studied and applied. MoS_2 can obviously improve the load-carrying capacity and friction-reducing properties of lubricating grease, but does not affect its oxidation stability and corrosion performance. Moreover, MoS_2 still possesses good lubricating property even in the lubricating grease containing impurities like dust. It was reported that the wear-resistance can be improved after MoS_2 was dispersed in the oil; especially in the high load, the wear-resistance effect was more obvious.

In addition, with the application and improvement of modern surface techniques, such as magnetic sputtering deposition, laser technology, ion assisted deposition, ion beam mixing and LB plating, the potential of MoS_2 in the lubrication area will be further excavated.

It was reported that the present application proportion of MoS_2 in various industrial departments was as follows: automobile (lubrication of components and fixers) 20%, aerospace industry (lubrication of vacuum radiation-resistance) 5%, industrial machine (common lubrication) 50%, composite material industry (lubrication for preparation of components) 15%, metallurgy industry (powder lubrication) 5% and other 5%.

1.3.3   WS$_2$

WS$_2$ can be applied as solid lubricant not only under the conventional conditions, but also under the harsh conditions, such as high temperature, high pressure, high vacuum, high load, radiation and in corrosive medium.

1.3.3.1   Crystal structure of WS$_2$

Similar to MoS$_2$ and graphite, WS$_2$ has close-packed hexagonal layered structure, as shown in Figure 1.10. The lattice constants are: a =0.318nm, c = 1.25nm. Each tungsten atom connects two sulfur atoms; tungsten atom and sulfur atom are connected by strong chemical bond, while sulfur atoms are connected by weak molecule bond.

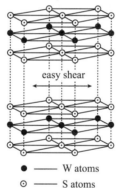

● —— W atoms

○ —— S atoms

Fig.1.10   Schematic diagram of crystal structure of WS$_2$

1.3.3.2   Properties of WS$_2$

The basic physical parameters of WS$_2$ are shown in Table 1.7.

Table 1.7   Basic physical parameters of WS$_2$

Items	Data
Molecular weight	248.02
Density/(g·cm^{-3})	7.4—7.5
Thermal stability	Begin to be oxidized at 425°C
Crystal type	Hexagonal lattice
Color	Gray and black with luster
Hardness (Mohs)	1—2
Friction coefficient	0.03—0.05

WS$_2$ is a chemical inert compound; it is hardly dissolved in any mediums, including water, oil, alkali and almost all kinds of acid; but it is sensitive to the free fluorine gas, hot sulfuric acid and hydrogen fluoride acid. It is innoxious material and does not corrode metals.

WS$_2$ has good heat resistance and oxidation resistance. Its heat resistance is better than that of MoS$_2$; the friction coefficient of WS$_2$ is very low in air until 500°C, and it is still low at higher temperature in the argon atmosphere. When the temperature is more than 400°C, WS$_2$ can be oxidized slowly to form compact WO$_3$ protective layer to restrain its further oxidization. WO$_3$ also possesses low friction coefficient, which can protect metal surface from adhesion. In the vacuum degree of 10^{-10}mmHg, WS$_2$ is stable until 1100°C; it will be decomposed when the temperature is greater than 1400°C. Moreover, the load-carrying capacity of WS$_2$ is also high; it is about three times that of MoS$_2$.

The cost of WS$_2$ is relatively high, about three times that of MoS$_2$. When the temperature is more than 400°C, WS$_2$ is suggested to be used in common atmosphere, while the cheaper MoS$_2$ should be used when the temperature is lower. WS$_2$ and MoS$_2$ exhibit almost the same property in vacuum, and both possess lubricity below 1320°C.

WS$_2$ can be used as additive of lubricating grease, paraffin and ozocerite; it can be also added to metal-base composite materials. The effect of WS$_2$ as friction moderator is better than that of MoS$_2$. Lubricating grease containing 5%WS$_2$ can save energy of 10% at low load, and save energy of 30% at high load and high speed conditions.

1.3.3.3 Preparation methods of WS$_2$

Natural WS$_2$ is extremely rare; therefore, the commonly used WS$_2$ is prepared by chemical methods. WS$_2$ used as lubricating material should have high-purity and micrograins. The fine grain size can be obtained conveniently using the colloid mill or gas stream crasher. Several preparation methods are introduced as follows.

1. Element synthesis method

This method is based on the principle of direct reaction of metal tungsten and sulfur (or sulfur vapor). The chemical formula is as follows:

$$W+2S=WS_2$$

The preparation conditions of WS$_2$ by this method are different. For instance, the mixture of a certain weight ratio of metal tungsten and sulfur powders is placed in the quartz tube; WS$_2$ can be obtained after the reaction for 24h at 800—900°C in nitrogen. The mixture of metal tungsten and sulfur powders can be also preheated at 110—250°C; then the reaction is conducted at 550—650°C; the free sulfur of obtained WS$_2$ powder with flaky crystal structure is removed by evaporation method; finally, WS$_2$ powder with purity of 99.9% and granularity of 0.1—2μm are obtained.

2. Tungsten trioxide method

Because WO$_3$ with high purity is easy to obtain, the preparation of WS$_2$ from WO$_3$ is interesting. One method is that the mixture of WO$_3$, S and KCO$_3$ is heated to 900°C in the condition of isolated air to conduct reaction; then the product of reaction is cooled, crushed and washed to obtain WS$_2$. However, the prepared product by this method always contains a small quantity of WS$_3$ or oxide of tungsten. Another method is that WO$_3$ is placed in the flowing H$_2$S gas, and WS$_2$ can be obtained after the reaction of 4h at 900—1200°C.

3. Ammonium tetrathiotungstate method

Ammonium tetrathiotungstate method is the most commonly used method to prepare WS_2 in industry at present. Tungstic acid and ammonia are mixed and heated to generate ammonium tungstate, then H_2S is injected to form ammonium tetrathiotungstate; the precipitated ammonium tetrathiotungstate crystallizes; and is roasted in the high temperature reverberatory furnace, finally WS_2 is obtained.

4. Ammonium tetrathiotungstate pyrolytic method

WS_2 nano-powders are prepared by ammonium tetrathiotungstate $((NH_4)_2WS_4 \cdot H_2O)$ pyrolytic process, the reaction formula is as follows:

$$(NH_4)_2WS_{4(s)} = WS_{2(s)} + 2NH_{3(g)} + H_2S_{(g)} + S_{(l)}$$

Ammonium tetrathiotungstate is raw material; the reaction temperature is 850—1040°C; reaction time is 1—2h; reaction gas is hydrogen or nitrogen, or their mixture.

5. Liquid deposition method

At room temperature, sodium sulfide reacts with sodium tungstate solution or ammonium tungstate to generate thiosulfate tungstate; When using the hydrochloric acid to acidify the solution to pH=2.5, thiosulfate tungstate decomposes and precipitates WS_3; as the impurity in the system can exist in a dissoluble state, a majority of impurity can be removed through filtrating and washing so as to obtain the further pure product; finally the precipitation is dried and heated to 170°C, and the termination product—WS_2 ultra-fine power with high purity is obtained. The main reaction formulas are as follows:

$$\left. \begin{array}{l} Na_2WO_4 + 4Na_2S + 4H_2O = Na_2WS_4 + 8NaOH \\ (NH_4)_2WO_4 + 4Na_2S + 4H_2O = (NH_4)_2WS_4 + 8NaOH \end{array} \right\} \quad (1)$$

$$\left. \begin{array}{l} Na_2WS_4 + 2HCl = WS_3 \downarrow + H_2S + 2NaCl \\ (NH_4)_2WS_4 + 2HCl = WS_3 \downarrow + H_2S + 2NH_4Cl \end{array} \right\} \quad (2)$$

$$WS_3 \xrightarrow{170°C} WS_2 + S \quad (3)$$

1.3.3.4 Preparation methods of WS_2 solid lubricating film

1. Magnetic sputtering method

The working gas was the mixed gas of argon (99.995%) and oxygen (99.5%). During deposition, it did not need heating and cooling. The target was WS_2. Before the gas was injected, the vacuum degree in the furnace was $10^{-5}Pa$; the total pressure of argon and oxygen was 0.5Pa or 1Pa. The sputtering power was 25W [19].

2. Chemical vapor deposition method (CVD)

Chemical vapor deposition is a technique of using the reactions between gas phases or gas and solid phases to generate solid sediment on the substrate surface. The chemical reactions of CVD include mainly two types: One is that the sediment is produced through the reactions of two gases or among several gases, e.g., the preparation of ultra-pure multicrystal silicon and nano-material (TiO_2). Another is that the reaction between a gas and solid substrate surface is conducted to deposit a thin film on the substrate, e.g., the preparation of integrate circuit, SiC container and parts with diamond film. The main characteristics of CVD method are as follows: the spreadability is good; the film can be deposited on the deep hole, staircase, depressed surface or three-dimensional object with complicated form and structure; the stoichiometric proportion of film can be controlled in a wide range; various crystalline and amorphous metals with high purity and required properties, semiconductors, and compound films can be prepared; the cost is low, and it is applicable for batch production; meanwhile, this method has good consistency with other processes. The shortcomings are that the reaction should be conducted at high temperature; the substrate temperature is high; the deposition rate is low; the gases initially taking part in the reaction and remain gases after reaction all have certain toxicity. Therefore, the application of CVD process is less extensive than that of sputtering and ion plating.

3. Electrodepositing method

Electrodepositing method was employed to prepare polycrystal WS_2 film on the conducting glass plate [20]. Before deposition, the substrate was cleaned by acetone. Different current densities ranging from $20mA/cm^2$ to $60mA/cm^2$ were used and the temperature of deposition bath was kept at $40°C$, $60°C$ and $80°C$; pH value was kept in the range from 7.0 to 9.5; finally WS_2 film was annealed for 60min at $300°C$ in vacuum.

1.3.3.5 Preparation methods of IF-WS$_2$

The studies on the preparation of IF nano-compound is one of research hotspots in recent years. Many countries have successfully prepared IF nano-compounds by various methods; the main applied methods are as follows [21].

1. Gas-solid or gas reaction synthesis method

Nano-scale transition metal oxide with stick or spheric shape was used as precursor; the nano IF transition metal sulfide with tube shape or onion structure can be synthesized through the gas-solid reaction. The vaporizing temperature of precursor WO_3 of transition metal sulfide was about $1400°C$. The synthesized IF-WS_2 by this technique possessed even shape and grain size and high purity.

2. Fluidized bed method

Fluidized bed method was used to prepare WS_2 nano-compound with onion-shape structure. The process of this method was that tungsten oxide nano-particles should be prepared at first; then nitrogen was employed to introduce the above nano-particles in the fluidized bed reactor; the nano-particles were mixed with the mixture of N_2 and H_2S, or mixture of N_2 and H_2; sulfuration or reduction reaction occurred at 600—$800°C$; nano WS_2 was obtained finally [22].

3. Microwave irradiation method

Germany first applied this method to prepare WS_2 nano crystal particles successfully. $W(CO)_6$ was first mixed with H_2S in hydrogen atmosphere; microwave oven with power of 0.915GHz or 2.45GHz was used to supply the energy required of reaction. The parameters for 0.915GHz oven were that the pressure was 30MPa; temperature was $260°C$ or $580°C$; the gas stagnation time was 8ms or 4ms. When 2.45GHz microwave oven was used, the parameters were that the pressure was 10MPa; the temperature was $160°C$; the gas stagnation time was 1.5ms.

4. Template technology

Template technology has extensively been studied and applied for the synthesis of carbon nanotube; the advantage of this technique is that the length and tube diameter are homogenous and can be controlled. With the template of CTAB, $(CTA^+)WS_4^{2-}$ flake was prepared by hydrothermal synthesis; $(CTA^+)WS_4^{2-}$ flake as precursor was calcined at high temperature; $(CTA^+)WS_4^{2-}$ flake was thermal decomposed and curled to form WS_2 nanotube.

1.3.3.6   Application of WS$_2$

The application of WS_2 can be classified as the following types.

1. WS$_2$ powder

WS_2 powder is the basic substance to prepare lubricating materials containing WS_2; it can also be directly used on the friction surface. For instance, WS_2 powder is smeared on the surface of screw chaser of giant bolt and surface of valve core of glass valve, or conveyed on the friction surface by airflow. Besides pure WS_2 powder, there are also mixed type powders containing WS_2 and MoS_2.

2. WS$_2$ grease

WS_2 powder as extreme pressure additive can augment application range of grease. For instance, composite calcium grease containing 1.5% WS_2 has been applied to bearings of high temperature tunnel kiln cars in the ceramic factory and steel plant and solved the problem that tunnel kiln cars can not be pushed off the tunnel due to the bearing lubrication failure caused by high temperature. The additive quantity of WS_2 varies with the kind and application occasion of lubricating grease; in general, the additive quantity of 1.5%—5% WS_2 is enough.

3. WS$_2$ oil agent

WS_2 powder, anticorrosive agent, suspending agent and engine oil can be made up an oil agent, which can be used for mechanical processing, such as metal cutting, drilling and stamping, and can improve workpiece smoothness, prolong service life of cutting tools and dies. For instance, when a bearing factory flanged the bearing sets on the 250 ton punch, they sprayed spindle oil containing 10% WS_2 on the die surface, the service life of die was increased by more than ten times.

1.3.4   ZnS

1.3.4.1   The crystal structure of zinc sulfide

The crystal structure of ZnS possesses sphalerite type (cubic structure, β phase) and wurtzite type (hexagonal structure, α phase). The used ZnS powder has typical hexagonal wurtzite structure, while the structure of sputtered film can be single hexagonal wurtzite structure, or mixed structure of hexagonal sphalerite and hexagonal wurtzite, exhibiting two-dimensional layered structure grown along the direction of C axis according to different sputtering power.

1.3.4.2   The properties of zinc sulfide

ZnS has a variety of excellent properties and can extensively be applied to many fields.

ZnS is white powder solid; as it is opaque and can not be dissolved in water, organic solvent, weak acid and weak base, it becomes an important pigment in the paint.

ZnS is easy to disperse and difficult to aggregate; it is neutral white and has good optical property; it is commonly used as the component of thermosetting and thermoplastic plastics, strengthening fiber glass, flame retardant, artificial rubber and dispersant agent.

ZnS is a kind of infrared optical material having high infrared transmittance in the wave band of 3—5μm and 8—12μm, and possesses excellent comprehensive properties of light, mechanic and heat.

ZnS nano-particles as oil additive have good lubricating property and can bear high load.

1.3.4.3   The preparation methods of zinc sulfide

ZnS nano-particles are generally used as oil additive to improve the lubricating effect. There are many methods to prepare nano ZnS; they can be classified as solid phase method, liquid phase method and gas phase method. The process of solid phase method is simple, the productivity is high and the obtained particles are relatively stable. Liquid phase method has relatively low demands to the reaction device; the prepared particles disperse evenly and have high purity. The particles with spheric shape, even distribution can be easily obtained by gas method; meanwhile, the particles are difficult to agglomerate, but the reaction involves high temperature, high pressure and has high demands to device.

1. Solid phase method

Solid phase method is using the proper zinc salt and sulfide to be grinded and mixed directly; the solid phase chemical reaction takes place under grinding and other mechanical actions so as to obtain ZnS. This method has many prominent characteristics: the manipulation is convenient; the synthetic process is simple; the conversion percent is high; the grain size is even and can be controlled; there is little pollution; this method can avoid or reduce the agglomeration phenomenon, which easily appears in the liquid phase method, and this phenomenon caused by intermediate step and high-temperature reaction [23].

Dihydrate zinc acetate and $Na_2S \cdot 9H_2O$ are used as reactants; sodium chloride and ethanediol are used as dispersant. ZnS with grain size of 20nm can be synthesized after admixture, grinding, reaction, washing, drying and calcination, etc.

$ZnCl_2$ and CaS are used as raw materials; through solid phase reaction the aggregate containing 12nm ZnS microcrystallite can be obtained; the size of aggregate is about 500nm. If 71% (volume fraction) $CaCl_2$ is added and used as diluting agent of reactant, dispersive ZnS particles with grain size of 16nm can be obtained. If 10—50nm CaS are used as raw material, through solid phase reaction between CaS and $ZnCl_2$, dispersive ZnS particles with grain size of 7—9nm can be got [24].

2. Liquid phase method

Liquid phase method includes mainly microemulsion method, emulsion method, homogeneous precipitation method, sol-gel method and hydrothermal method, etc.

(1) Microemulsion method.

Since 1982, the preparation of metal clusters particles of Pt, Pd, and Rh in the W/O microemulsion exploited a new method to prepare nano-particles. Compared with traditional preparation methods, it has obvious advantages. Microemulsion is a thermodynamic stable, isotropic, hyaloid or semitransparent dispersion system composed of two or more kinds of immiscible liquid. Actually, it is constituted by the liquid microdrops stabilized by the interfacial film of surface active agent, the size of microdrop ranges from 10nm to 100nm. According to the difference of continuous phase of microemulsion, this method can be classified as oil-in-water type (O/W), water in oil type (W/O) and bicontinuous phase type. In the special condition, microemulsion can also be formed by polar organic substance, water and oil; the formation is related with its composition only, and unrelated with preparation method. Reverse micelle means that surface active agent is dissolved in the organic solvent; when the concentration exceeds critical micelle concentration, the liquid particle structure with hydrophilic polar head inward and hydrophobic chain outward is formed. The water kernel of microemulsion is a "microreactor" with big interface, which can solve various compounds. The preparation of ZnS by microemulsion method is just making use of the ion-exchange reaction occurred in the "microreactor" of microemulsion. If Zn^{2+} exists in the water, and sulfur source (like CS_2) exists in the organic phase (like benzene), Zn^{2+} can slowly react with S^{2-} to generate ZnS through the movement and exchange of interface ions.

In the supercritical CO_2, two kinds of microemulsion containing Zn^{2+} and S^{2-} respectively are mixed; the water drop in the CO_2 acts as nanoreactor; then ZnS hollow ball can be obtained.

(2) Emulsion method [25].

Emulsion is a system of one phase (like water) in the form of microdrops dispersed in another phase (like oil) of two kinds of immiscible liquid under the existence of a certain quantity of emulsifier. Differing from microemulsion, the emulsion is thermodynamic unstable system; the volume of dispersed microdrops is relatively large. Continuous oil phase can disperse the water phase containing reactant to minute drops; the chemical reaction is taking place in the innumerable minute drops. As the drops have small volume and are fixed, the generated grain size in the drops is controlled; the final grain size of the product particle depends on the volume of minute drops. After the formation of superfine particles, they are washed to remove emulsifier, and then dried to obtain superfine particle sample.

(3) Homogeneous precipitation method.

During the precipitation, two processes of deposition and solution exist. However, if the concentration of precipitating agent in the solution is controlled and increased slowly, the deposition in the solution can achieve or close to the balance state. In this process, the sediment can homogeneously be precipitated from the solution. This method is called as homogeneous precipitation method.

During using this method, the chemical reaction makes the crystal ions released from the solution slowly and homogeneously. The added precipitant does not induce the precipitation reaction immediately in the solution; it is slowly hydrolyzed under the heating condition of precipitant and it reacts homogeneously in the solution. Nano-particles have undergone the nucleation and growth processes, when they precipitate from the liquid phase. When nucleation rate is lower than growth rate, it is helpful to obtain large and a small quantity of coarse particles; when nucleation rate is larger than growth rate, it is beneficial to form nano-particles. Therefore, in order to obtain nano-particles, the nucleation rate should be larger than growth rate.

The soluble salt of zinc and sulfur sources (NaS, NaHS, $(NH_4)_2S$, $Na_2S_2O_3$, TAA and $SC(NH_2)_2$, etc.) are mixed to precipitate in very dilute water solution and obtain micron scale and submicron scale powders with spheric shapes and good dispersion; the grain size can be controlled through adjusting the concentration and pH value of reaction solution.

Zinc acetate and sodium thiosulfate solution are mixed; then a small quantity of acetic acid is added to adjust the pH value; meanwhile, the temperature should be controlled; ZnS ultra-fine powders with grain size of 200nm and homogeneous dispersion can be obtained. The effect is optimal when the pH value keeps around 3.0, it is proper when the reaction temperature keeps around 80°C.

Thioacetamide (TAA) is used as sulfur source; TAA water solution hydrolyzes under the condition of acidity and a certain temperature to release H_2S homogeneously; zinc sulfate is adopted as zinc source; nano ZnS with average grain size of 40—50nm can be synthesized.

(4) Sol-gel method.

Sol-gel method is that the metal organic or inorganic compounds are used as raw material; after the chemical reactions of hydrolysis and condensation, submicron superfine grain sol with dispersive liquidity can be formed in the generated solution; then the sol combines with superfine grains to form gel.

Sol-gel method has many characteristics: the product has good evenness; the calcination temperature is lower than that of high temperature solid phase reaction; the product purity is high; the banded emission peak is narrow; the relative luminous intensity and relative quantum efficiency of luminous body can be improved.

3. Gas phase method

Gas phase method is using gas or making substance converting to gas to process the physical or chemical reactions in the gas state; nano-particles are formed after condensation, coacervation and growing in the gas protective atmosphere. The prepared nano-particles by this method have many advantages, such as high purity, small particle size, little agglomeration and controllable component.

(1) Chemical gas phase deposition method.

Gas phase deposition method is a commonly used method to prepare nano-particles. The raw material is heated to evaporate in the inert atmosphere, such as He and Ar under low pressure; the evaporated atoms or molecules lose their kinetic energy under the impact of inert gas atoms to aggregate nano crystal grains with a certain size.

(2) Spray pyrolysis method.

Spray pyrolysis is the most effective and most commonly applied method to prepare spheric lighting powders. The main process is generally divided into two steps: The first step is the evaporation of liquid drops from the surface. With the evaporation of solvent, it achieves hypersaturated state; minute solid phase gradually precipitates from the bottom of the drop, then gradually expand to the surrounding region of the drop and finally cover the whole surface of the drop to form a layer of solid phase shell. The second step is to dry the drops. This step is more complicated, including the formation of pores, rupture, expansion, shrinkage and the growth of grain "hair". Spray pyrolysis is also a method to prepare ultra-fine particles.

1.3.4.4   The preparation methods of ZnS film

1. Electric brush plating method

Electric brush plating is a special mode of electric plating; the plating layer can be obtained by using a plating pen, which is supplied continuously with electrolyte and connected with anode, to wipe the workpiece surface connected with cathode. Compared with conventional electric plating, brush plating possesses the following characteristics: ① The equipment is simple; it is convenient to carry; it does not need large plating bath. ② The process is simple; it is convenient to manipulate; the surface, where the plating pen can touch, all can be plated; especially this method is applicable for field maintenance of machine parts, which are difficultly disassembled. ③ Plating layers are of a great variety, the bonding strength is strong between the layer and substrate; the mechanical properties are good; it can meet the requirement of various maintenance. ④ The deposition speed and productivity are high, but high current density must be adopted to operate. ⑤ The brush plating solution does not contain cyanide and highly toxic constituent, it is safe to manipulate and has little environmental pollution.

ZnS film had been prepared on the substrate of Ti and conductive glass by electric brush plating method; the current density was $80mA \cdot cm^{-2}$; the film was deposited under different temperature ranging from $30°C$ to $80°C$. The thickness of ZnS film prepared under low current density was less than $0.5\mu m$, while the film prepared under high current density showed coarse grain size, but the film thickness did not increase [26].

2. Chemical bath deposition method [27]

Chemical bath deposition (CBD) method is a process of using the controllable chemical reaction to deposit ZnS film. In most experimental methods, substrate is immersed in the solution containing chalcogenide oxide and metal positive ions; complexing agent is used to control the hydrolyzation of positive ions. This technique depends on the release of sulfur ions in the solution; the metal ions in the solution should be reduced to a low concentration state; when ion recombination velocity is more than solution rate, the film will be produced. Chemical bath deposition method is generally used to prepare ZnS film in the alkaline solution adopting the hydrazine hydrate (N_2H_4) as coordination agent.

Hydrazine hydrate has high toxicity; so acidic solution can be adopted to prepare ZnS film. Figure 1.11 shows the schematic diagram of water bath experimental device. According to the required mixture proportion, the analytically pure $ZnCl_2$, urea and thioacetamide are added to 200mL de-ionized water successively; the pH value of the mixed solution is adjusted to 4 by hydrochloric acid; then beaker A is placed in the beaker B; the solution is well agitated and heated when the solution becomes clear. After the solution achieves the required temperature, the substrate is placed in the solution to deposit film. This method was employed to deposit ZnS film on the glass coated with SnO_2. The water bath solution contains 0.315g $ZnCl_2$, 0.162g thiourea $(CS(NH_2)_2)$, 5.44mL ammonia solution of 28% and 3.34mL hydrazine hydrate.

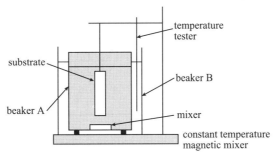

Fig.1.11 Schematic map of experimental device [27]

3. Sputtering method

Sputtering method has extensively been applied to prepare thin films. The ZnS film prepared by radio-frequency magnetic sputtering in the pure Ar atmosphere shows relatively complete and homogeneous structure; the higher the sputtering power, the smaller the crystal grains.

4. Chemical vapor deposition method

Chemical vapor deposition method was developed since the 1980s; compared with the sputtering method, it had many advantages: the deposition temperature is low; deposition rate is large; the doped concentration can be easily controlled; the deposited film has high quality and it is easy to realize the large scale production. The ZnS/Mn film prepared by low pressure chemical vapor deposition method has a hexagonal crystal structure with the preferred orientation of (00.1).

5. Pulsed laser deposition method

Since the end of 1980s the pulsed laser deposition method was developed rapidly for preparing the thin films. Its prominent advantage is that the prepared film has almost the similar chemical composition to target; this point simplifies the control of the composition of film. Therefore, it is applicable to prepare the films with complex composition and high melting-point.

Using YAG solid laser (1064nm) and XeCl (308nm) excimer laser to investigate the effect of energy densities of different pulsed laser and different wave lengths on the ZnS film, it was found that within the same deposition time, the ZnS film prepared by low pulsed energy density exhibited dense crystal structure and good quality; the lightening brightness was also improved [28]. Under the same laser power, the laser wave length of YAG solid laser was relatively long, while that of XeCl excimer laser was short. The deposited film by the latter method was smoother; the particle size reached several tens nanometers, or even less.

6. Reactive sputtering + sulfuration

As the kinetic energy of sputtered material particles (1—10eV) is far more than that of evaporated material particles (−0.1eV), the former particles have higher migration rate on the substrate surface so as to form a dense film. The materials as target cathode can be metal, alloy, low valent metal compound or high-doping semiconductor. During the process of sputtering and migrating of material particles, they collide with the reactive gas molecules to generate compound—medium film. Therefore, it is called reactive magnetic sputtering technique.

The reactive sputtering method was used to deposit ZnO film on the substrate of glass and quartz firstly; then it was annealed under different conditions and sulfurized in the H_2S atmosphere to obtain ZnS film finally [29]. The thickness of the ZnS film was about 200nm. The result showed that the ZnO film annealed in air and pure nitrogen could be converted to ZnS with hexagonal structure entirely after the ZnO film was sulfurized for 2h; but only part of ZnO film was converted to ZnS when it was annealed in the pure oxygen and vacuum.

1.4   Inorganic Solid Lubricant

The most commonly used inorganic solid lubricants are graphite and fluoride graphite. Although the lubricating properties of talc, mica and silicon nitride are poor, their electric insulating properties are good; they can be used as solid lubricants and lubricating fillers under high temperature and special conditions. Boron nitride has the same layer structure and similar characters to graphite; it is white powder and can be used as high temperature and insulating heat-barrier lubricating materials.

1.4.1   Graphite

The appearance of graphite is black and has aliphatic soapy feeling. Graphite has apparent hexagonal layer structure. Its density is 2.2—2.3g/cm^3; its melting point is 3527°C. The molecular structure of graphite makes the carbon atoms in the same layer firmly bond together; the bonding is difficult to be destroyed. However, the bonding strength between layers is weak; the layers are easy to slip under the action of shearing force. In general, the friction coefficient of graphite ranges from 0.05 to 0.19.

1.4.1.1   The crystal structure of graphite

Graphite is an isomeride of carbon; its crystal structure is shown in Figure 1.12. In the identical plane; each carbon atom connects with three adjacent carbon atoms by covalent bond emerging 120° angle. The distance between carbon atoms is 0.142nm. The carbon atoms between layers are connected by weak molecular force; the distance between layers is 0.3335nm. Each carbon parallel layer presents in a tortoise-like shape, there are several thousands such layers in 1μm thick crystal.

Graphite has two types: natural graphite and artificial graphite. The natural graphite is crushed from mineral graphite by hammer crusher with high speed of 8000r/min after choiceness, flotation and dehydrogenation drying; then it is reclaimed through tornado separator and dust collector to form the minute particle graphite. The purity of graphite after choiceness and purification can reach 98%. Artificial graphite is calcinated from petroleum coke at 1400°C; after crushing, coal tar pitch and anthracite are added; then it is calcined at 800°C and soaked by coal tar pitch; subsequently, it is graphitized at 2600—3000°C and crushed again to a grain size of 0.5—250μm. The obtained graphite is black scaled powder with purity of 98.5%.

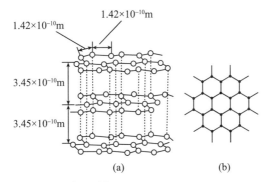

Fig.1.12   The crystal structure of graphite

1.4.1.2   Physical properties of graphite

The physical properties of graphite are shown in Table 1.8.

Table 1.8   Physical properties of graphite

Performance	Value
Relative atomic mass	12.01
Crystal type	Hexagonal crystal system
Density $(20°C)/(g \cdot cm^{-3})$	2.268
Melting point/$°C$	3527
Vapor pressure $(2204°C)/Pa$	0.1
Specific heat capacity/(kJ/(kg$\cdot°C$))	8.5
Specific resistance $(20°C)/(\Omega \cdot m)$	4×10^{-7} (Parallel to the base plane) 4×10^{-5} (Vertical to the base plane)
Coefficient of heat conductivity $(20°C)/(W/(m\cdot°C))$	400 (Parallel to the base plane) 6 (Vertical to the base plane)
Coefficient of thermal expansion/(cm/(cm$\cdot°C$))	-0.36×10^{-6} (Parallel to the base plane) 30×10^{-6} (Vertical to the base plane)

If graphite is heated in the atmosphere with oxygen, it can be oxidized to carbon dioxide, when the temperature is more than 325°C. Thus, antioxidant should be added if it is used at high temperature for a long time. If a proper anti-oxidation method is selected, the bearing with wear life of several thousand hours may be obtained in air at 675°C.

Graphite has good high temperature stability; it is a substance that the higher the temperature, the bigger the strength. From room temperature to 2500°C, the tensile strength, bending strength and compressive strength all increase with the rise of temperature; the maximum value can reach two times as high as that at room temperature. Figure 1.13 shows the relation of tensile strength with temperature of graphite. The shearing strength also has such characteristic. It can be inferred that the lubricating property of graphite will also decrease at higher temperature.

Graphite is not molten even at 3200°C in vacuum; so from this point, using graphite as lubricant in vacuum should be more appropriate than in air. But in actual, the lubricating effect of graphite is impacted by absorbed gas; once the absorbed gas is eliminated, graphite will lose its lubricating effect. Thus, graphite can not be used as lubricant in vacuum. The highest temperature can reach 1000°C, when it is used in air for a short time.

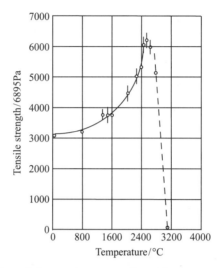

Fig.1.13   The tensile strength-temperature curve of graphite

Graphite has good adhesivity and is good conductor of heat and electric; it has low evaporability in vacuum. Therefore, it can be applied to the lubrication of aerospace and other special equipment.

Graphite has good chemical stability, it is indissoluble in chemicals and solvents; it is non-toxic. Meanwhile, graphite has excellent radioresistance; it has strong α-ray and neutron-ray resistance property; even if it is subject to strong radiation by neutron ray of $10^{20}/cm^2$, no changes can be detected.

Another characteristic is that graphite and water can be concurrent; even if water is used as coolant, the lubricating effect of graphite does not get worse, which is different from that of MoS_2. The colloid graphite with good dispersion in water has been merchandized.

1.4.2   BN

1.4.2.1   The crystal structure of BN

The crystal structure of BN includes hexagonal crystal system, rhombohedral crystal system, cubic crystal system (the hardness is similar to diamond; it can be used as abrasive material) and turbostratic structure, etc. The BN used as solid lubricant belongs to hexagonal crystal system. The crystallization of BN with hexagonal structure is similar to that of graphite. The crystal structure of BN is shown in Figure 1.14.

Boron oxide or borax is reduced and nitrided in the ammonia containing porous medium or foaming medium filling at 800—1200°C to obtain BN with turbostratic structure; then it is heated to more than 1800°C and converted to hexagonal crystal system.

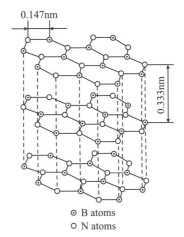

0.147nm

0.333nm

⊙ B atoms
○ N atoms

Fig.1.14　Crystal structure of hexagonal BN

1.4.2.2　The physical and chemical properties of BN

BN with high purity is white or faint yellow micro-powder. It has good heat resistance, chemical stability and electric insulation property. It is still stable and can maintain these properties in the inert gas at 2800°C. Thus, it is suitable to be used as high-temperature lubricant, friction-reducing heat insulation material or filler of special resin.

Many physical properties of BN are similar to those of graphite; so BN is called as "white graphite". BN, mica, talcum powder, silicate, and fatty acid are called by a general name—white solid lubricant. The physical and mechanical properties of BN are listed in Table 1.9.

Table 1.9　Physical and mechanical properties of BN

Property	Value
Density/(g·cm^{-3})	2.15—2.2
Tensile strength/MPa	2.5 (1000°C)
Compression strength/MPa	Parallel 315 (room temperature) Vertical 238 (room temperature)
Bend strength/MPa	Parallel 111 (room temperature) Vertical 51 (room temperature) Parallel 15 (1000°C) Vertical 7.5 (1000°C)
Friction coefficient	0.2—0.3
Hardness	2 (Mohs)
Coefficient of linear expansion (20—600°C)/(10^{-6} · °C^{-1})	Parallel 8.58 Vertical 6.33
Coefficient of heat conductivity/(W·m^{-1} · °C^{-1})	Parallel 15.16 (300°C) Vertical 28.76 (300°C) Parallel 12.35 (1000°C) Vertical 26.67 (1000°C)
Electrical resistivity/(Ω·cm)	1.7×10^{13} (25°C) 3.1×10^{4} (1000°C)
Maximum working temperature/°C	In air 1100—1400°C In nitrogen 3000°C

1.4.2.3   Usage of BN

When used as lubricant, BN can be dispersed in the lubricating oil or grease, water or solvent and sprayed on friction surface to form dry film. BN can be filled into surface layers of resin, ceram, and metal to act as heat resisting high temperature self-lubricating material applied to astronavigation engineering. BN powder can be also directly smeared on the surface of guideways.

BN suspended oil is white or faint yellow; it is mostly applied to the lubrication of the synthetic fiber textile machines. The properties and applications of BN lubricating oil are given in Table 1.10.

Table 1.10   Properties and applications of BN lubricating oil

Base oil	BN content /%(mass)	Application	Feature	Remark
Silicone oil (Methyl phenyl silicone oil, etc)	1—20	Lubricant used in high temperature (used in synthetic fiber industry)	Good heat resistance, white, can be used above the temperature of 2500°C	Can be used combining with 1%—15% lithium soap
Silicone ester	5	—	Increase the heat resistance of graphite oil	Can be used combining with graphite
Silicone oil Polyalkylene glycol Polyphenylene oxide Ester oil Fossil oil Fatty acid calcium grease	2—20	Grease used in hyperthermal environment (airplane, rocket, missile)	—	0.1%—10% antioxidant and stabilizer can be added
Fluorinated ethylene propylene copolymer	40—65	Grease used in high temperature	Stabilize at 315°C	—
Fluorinated hydrocarbon (Fluorine oil)	0.5—0.75	Heat resisting compressor oil	—	—
Silicone oil Polyphenylene oxide Fluorine oil	15—45	Grease used in high temperature	—	—

The application examples of self-lubricating material containing BN as additive are shown in Table 1.11.

Table 1.11   Resin composite materials containing BN and their applications

Self-lubricating material	BN content /%(mass)	Application	Feature	Remark
Silicone resin	42		Heat resisting until the temperature of 650°C	Sintered on substrate surface to be used
Bakelite	54	Guide material		
Fluororesin	5—25			
PTFE	10	Guide and bearing materials	—	—
PTFE	30	Guide material	—	Combined use with the Ag powder
Polyamide Polyimide Poly-glyoxaline	1—70	Guide material	—	—
Fe—Cu—Pb—Ni alloy	0.5—5	Shelf-type pantograph collector material	Good lubricating and arc resistance properties	—
Aluminum oxide (Al_2O_3)	2—7	Bearing/cableway used in the high temperature	—	—

1.5   Organic Solid Lubricant

Various high molecular materials all can be used as solid lubricant, such as wax (paraffin wax, mineral wax, bees wax and halogen wax, etc.), solid fatty acid, alcohol, biphenyl, paint (indanthrene and phthalocyanine, etc.), thermoplastic resin (polytetrafluoroethylene, polyethylene, nylon, polyformaldehyde, and polyphenyl thioether, etc.), and thermosetting resin (phenolic aldehyde, epoxy, organic silicon, and polyurethane, etc.)

High molecular material can be added to other lubricants in the form of powder; it can be also added to other solid lubricant (like molybdenum disulfide.) and used as matrix of high molecular base composite lubricating material.

Organic molybdenum compound is composed of Mo—P—S compound and Mo—C—S compound, such as dialkyl disulfide molybdenum phosphate, (normal-butyl isooctyl) dithiocar phosphate sulfate molybdenum oxide, and dialkyl disulfide amino molybdenum formate. They all belong to oil-soluble organic molybdenum; when they are added to lubricating oil and grease as friction moderator, reaction can occur on friction surface under a certain temperature and pressure to generate molybdenum disulfide.

Melamine cyanurate agent (MCA) is a new kind of high molecular organic compound. It has various properties of solid lubricant; it can be used in the forms of powder, thin film and composite material.

The basic properties of commonly used organic solid lubricants are shown in Table 1.12.

Table 1.12 Basic properties of several kinds of organic solid lubricants

Item		Material	Density /(g/cm³)	Tensile strength/MPa	Compressive strength/MPa	Impact strength /((J/m²)·10⁻²)	Coefficient of heat conductivity /(W/(m·K))	Coefficient of linear expansion /(10⁻⁶/°C)	Commonly used temperature extremes/°C
		PE (high density)	0.96	20—31	19—25	2.8—11	0.12	110—130	110
Thermoplastic macromolecular material		Nylon	1.09—1.14	49—84	35—91	1.1—2.2	0.21—0.24	99—103	<100
		Polyacetal	1.43	70	127	1.7—2.2	0.05—0.22	81	<100
		PTFE	2.2	9	4	0.96—1.5	0.24	99	260
		Polyimide	1.41—1.43	94	244	2.8	0.31	54	<260
Thermosetting macromolecular material		Bakelite	1.3—1.4	50—70	200—270	1.8—2.8	—	40—60	<100
		Epikote	1.2	35—84	—	1.7—5.0	0.19	50—90	<100

1.5.1   Polytetrafluoroethylene

Polytetrafluoroethylene (PTFE) is fluoric organic high molecular compound; it was successfully developed by American Dupont Company in 1938. Since the second world war, PTFE was mainly applied to military industry; it began to be applied to civilian industry from 1950. The molecular formula of polytetrafluoroethylene is as follows:

$$
\begin{array}{cccccc}
\text{F} & \text{F} & \text{F} & \text{F} & \text{F} & \text{F} \\
| & | & | & | & | & | \\
-\text{C}-\text{C}-\text{C}-\text{C}-\text{C}-\text{C}- \\
| & | & | & | & | & | \\
\text{F} & \text{F} & \text{F} & \text{F} & \text{F} & \text{F}
\end{array}
$$

The bonding energy between C—F in the PTFE can reach 460kJ/mol (the bonding energy between C—H is 365kJ/mol); this high bonding strength makes PTFE very stable in the organic compounds. In the molecular structure of PTFE polymer, molecular chain shows spiral shape and is divided into two layers: the inner layer is carbon chain, while the outer layer is fluorine atom layer; the latter densely wraps the carbon chain. Because fluorine atom has the highest electronegativity in all elements, and it is charged negatively, so the positive charge of carbon atom is shielded; meanwhile, the repulsion exists between the negative charges of adjacent fluorine atoms, which leads to extremely low cohesion and absence of branch of molecular chain, it is basically a linear chain composed of $(CF_2—CF_2)_2$. These intrinsic factors make PTFE possess many excellent characteristics: It can resist coldness and hotness; it can work at temperature ranging from $-100°C$ to $280°C$ for a long time; the surface energy is extremely low (18.5×10^{-3}N/m, $\theta = 112°C$); it is difficult to be moistened; it possesses excellent drainage, oil drain and non-viscous properties; it is hardly corroded by chemicals; another important property is its prominent low friction characteristic.

The structure of PTFE polymer is amorphous with banded structure composed of 150—900Å thick tabular grains. Therefore, the deformation and rupture occur easily. The X-ray diffraction analysis also indicates that PTFE molecules are surrounded by outer-shell electrons with smooth distribution; the molecules exhibit columnar streamlined structure; the interaction between molecules is small, and the attractive power is weak. These structure characteristics essentially determine the easy slippage ability of PTFE. It has a very low friction coefficient, $\mu_{static} = \mu_{kinetic} = 0.04$. That is the lowest friction coefficient among all the materials at present, meanwhile, it has the highest chemical stability among all present plastics. PTFE has excellent self-lubricating friction-reducing property, which does not change in a wide temperature range; it can also keep lubricating property even in vacuum; its lubricating property is little impacted by the environment. However, its wear-resistance is poor; it is easily worn off under heavy load. Thus, PTFE and hard metal are usually used through combination, which can not only sufficiently play the role of friction-reduction, but also effectively improve its wear-resistance.

PTFE has been honored as king of plastics, it is a good material for preparing the products with high performance requirement. Its advantages are as follows.

(1) Wide application range for high and low temperature usage.

It can be used continuously for a long time in a range of -200—$260°C$; when the temperature is more than $200°C$, microcontent of PTFE begins to decompose; but it begins to decompose significantly at $400°C$; decomposes rapidly at $415°C$; it sublimes at $450°C$.

(2) Excellent chemical stability.

PTFE has the best chemical stability in engineering plastics. Even in the high temperature environment, it does not act with concentrated acid, diluted acid, concentrated alkali and strong oxidant; it is not corroded even in the boiling aqua regia; it does not react with most organic solvents, such as alcohol, ether and ketone, etc.; it can not be swelled by water or oil.

(3) Good electric insulativity.

Its electric insulativity is not affected by the environmental condition, working temperature, moisture and frequency.

(4) Prominent surface non-viscous performance.

As PTFE has extremely low surface energy, almost all adhesive materials can not adhere to its surface.

(5) Extremely low friction coefficient.

PTFE, graphite and MoS_2 are regarded as three most important solid lubricants, among them, PTFE has the lowest friction coefficient to 0.03, it is worthy to notice that its kinetic friction coefficient is much higher than its static friction coefficient, the former is 0.12, while the latter is 0.03, and it increases with the rise of sliding speed, so the creep phenomenon does not occur even at extremely low sliding speed.

(6) Good wear-resistance property.

Although pure PTFE has poor wear-resistance, various fillers as the reinforcing agent are usually added to PTFE, in consequence, its wear-resistance can be significantly improved; even it has better wear-resistance than other materials (such as cast iron and copper).

(7) Resistance to ageing in air.

When PTFE is exposed in air for a long time, its surface and various properties keep constant.

(8) Completely incombustible.

(9) PTFE has low hardness, soft texture; and the property to absorb and contain foreign bodies. It can easily transfer to the metal surface; meanwhile, it is easily to be machined.

Its limitations and shortcomings are as follows.

(1) It has big linear expansion coefficient ranging from $8×10^{-5}/°C$ to $25×10^{-5}/°C$; the value is 10—20 times larger than that of common metal.

(2) It has poor heat conductivity $0.24W/(m·°C)$, which is only 1/300 that of common metal.

(3) It has low strength.

(4) It has small elastic modulus.

(5) It is easy to creep.

After copper powder, graphite, MoS_2, glass fiber, and carbon fiber are added to PTFE, its main advantages can be kept, while its limitations and shortcomings are significantly improved. Therefore, the filled PTFE in engineering has more extensive application than pure PTFE.

Generally PTFE has two present types: one is white powders($\phi = 0.1$—5μm); the other is emulsion (solid content < 60%, $\phi < 1$μm). The composite solid lubricating films mainly include adhering type, composite plating type and TUFRAM type. Besides, there are also other types of solid lubricating films.

1.5.2   Polythene

Polythene (PE) is a high molecular compound formed through the polymerization of ethene. There are only two elements of C and H in the molecular structure of polythene; its molecular formula is $(CH_2$—$CH_2)_n$. According to different polymerization conditions, the actual molecular weight of product ranges from ten thousands to several millions. The most important structure factors influencing the performance of polythene are magnitude and distribution of molecular weight, molecular branching degree and crystallinity. In the low concentration range, the crystallinity is in direct proportion with concentration.

According to different polymerization processes, there are three types of polythene as follows.

(1) High pressure polyethylene (low density polyethylene, LDPE). Its polymerization pressure is 100—200MPa; the temperature is 180—200°C; the density is 0.910—0.925g/cm^2; the crystallinity is 55%—65%.

(2) Medium pressure polythene. Its polymerization pressure is 1.8—8MPa; the temperature is 130—270°C; the density is 0.926—0.940g/cm^2; the crystallinity is 90%.

(3) Low pressure polythene (high density polyethylene, HDPE). Its polymerization pressure is 1.4MPa; the temperature is 100°C; the density is 0.941—0.965g/cm^2; the crystallinity is 85%—90%.

As the molecular structure of polythene does not contain polar group, it has excellent dielectric property, low water absorbability and good chemical stability; it is indissoluble in any solvents at room temperature. However, it has poor stability in the hydrocarbons and oils; swelling or discoloration may be caused. A small amount of polythene can be dissolved in chlorinated hydrocarbon, naphthalene tetrahydride and decahydronaphthalene at high temperature (more than 70°C).

There are many branched-chains in the molecule of high pressure polyethylene; the crystallinity is low; the softening point is 105—120°C. Its mechanical strength, permeability, water vapor permeability and solvent resistance are inferior to those of medium pressure and low pressure polythenes, but its softness, extensibility, resistance to impact and transparency are all better than those of medium pressure and low pressure polythenes. There are less branched-chain in the molecule of medium pressure polyethylene; the crystallinity is high. Thus, it has best mechanical performance, high strength, good resistance to high temperature(the softening point is around 130°C), permeability and water vapor permeability. Low pressure polythene has hard texture.

Polyethylene has low friction coefficient. The static friction coefficient of low density polyethylene is 0.27—0.33, while its kinetic friction coefficient is 0.26—0.33.

The static friction coefficient of high density polyethylene is 0.12—0.18, while its kinetic friction coefficient is 0.08—0.11.

Low pressure polythene has a certain self-lubricating property and good wear-resistance. Under favorable lubricating condition, its wear-resistance is about one time higher than that of cast iron.

1.5.3 Nylon

Nylon, namely polyamide, the widely used nylon includes nylon 6, nylon 66, nylon 610, nylon 1010 and MC nylon.

The common characteristic of nylon material is high mechanical strength; its compressive strength is 60—90MPa; shearing strength is 40—60MPa; impact strength is 10—49J/cm^2; tensile-strength is 50—65MPa. It has obvious and high melting point; it does not become soften with the increase of temperature. For instance, the melting point of nylon 6 is 215°C; no obvious deformation can be observed at 150°C; it can be used below 120°C. The general products can be used at −45—100°C. Nylon has good toughness and corrosion resistance, it does not act with the solvents, such as weak base, alcohol, ester, ketone, gasoline, grease, oil and water; it has poor corrosion-resistance to strong acid only; it can be dissolved in hot concentrated sulfuric acid and phenol. Nylon has excellent wear-resistance, which is superior to that of copper and common steel; meanwhile, it possesses a certain self-lubricating property. It has low friction coefficient; the friction coefficient is 0.08—0.15 under oil lubrication condition when nylon rubs with metal, but the friction coefficient under dry condition is 0.34—0.37. Nylon also has many advantages, such as mould-resistance, non-toxicity and low cost. The shortcomings are big creeping and hydroscopic property.

Nylon has good friction characteristic. When PV value is 0.11MPa·m/s, it can work normally under unlubricated condition. While under lubrication condition, the PV value under normal temperature can reach 5.5. The prominent characteristic of nylon is big specific strength; it has the smallest density (1.11—1.14g/cm^3) among the commonly used wear resistant materials.

In summary, nylon is a good wear resistant material. Its shortcomings are strong hygroscopic property, poor creep-resistance and poor thermal stability, etc. They can be improved through adding different fillers.

1.5.4 Polyformaldehyde

Polyformaldehyde (POM) is classified as polyoxymethylene homopolymer and polyoxymethylene copolymer. The main chain of homopolymer is composed of continuous —C—O— bond, while the main chain of copolymer is composed of —C—O—C— bond and —C—C— bond. As —C—O— bond is shorter than —C—C— bond, homopolyformaldehyde has high density, crystallinity, melting point and strength; it is easy to decompose; it has poor heat stability; meanwhile, it has low stability to acid and alkali. Copolyformaldehyde has low density, crystallinity, melting point and strength; but it is not easy to decompose; it has good heat stability; it has good stability to acid and alkali.

Copolyformaldehyde possesses good mechanical properties; its compressive strength is 112MPa; tensile strength is 616MPa. It has excellent anti-fatigue property; under the circulating tensile and compressive load, the fatigue strength can achieve 35MPa. It can resist twist and restore rapidly; it can be completely recovered at once when the load is removed. Compared with nylon, copolyformaldehyde has high hardness, good rigidity, extremely small water absorption and good creep resistance; thus, its product has good dimensional stability. It also has good pharmacy-resistant property; it can resist aldehyde, ester, ether, hydrocarbon, weak acid and weak base, etc.; it has good petrol-resistance and lubricating oil resistance, but the UV-resistant performance is poor. In order to improve the weatherability, carbon black and ultraviolet ray absorbent should be added. Polyformaldehyde can work under a wide temperature range; the working temperature of homopolyformaldehyde is −40—149°C; that of copolyformaldehyde is −40—100°C; the hardness of copolyformaldehyde at 100°C is equivalent to that of polypropylene at 30°C.

Polyformaldehyde has self-lubricating property, high wear resistance and good friction characteristic. Compared with nylon, polyformaldehyde has lower friction coefficient (0.15—0.35); meanwhile, the static and kinetic friction coefficients are similar. Therefore, polyformaldehyde possesses good creeping resistance.

If PTFE is added to polyformaldehyde, the wear resistance can be improved. When rubbed with 1045 steel, its friction coefficient decreases from 0.44 to 0.21; the static and kinetic friction coefficients are close; under the low speed motion condition, no creeping phenomenon occurs under dry and oil lubrication conditions.

1.5.5   Phenol Formaldehyde Resin

Phenol formaldehyde (PF) resin is also called as bakelite. The resins obtained by the fasculation of phenolic compound and aldehyde compound are called by a general name, phenol formaldehyde resin. The phenol formaldehyde resin obtained by the fasculation of phenol and formaldehyde is most important. As the molecule structure of phenol contains polar groups of —OH and —CH, the adhesive force is strong. On the other hand, its structure contains many benzene rings, the associated density is large; thus, phenol possesses a certain mechanical strength; the heat resistance is also good; it can be used at the temperature over 100°C. When it is in the high temperature environment for a long time, its tensile strength, compressive strength and shearing strength all decrease slightly; it has excellent flame-resistance, good chemical corrosion- resistance, wet fastness, rigidity and small creeping; it is not easy to deform or cockle; the moulding is simple, and the cost is low.

However, phenol formaldehyde resin is brittle; its alkali-resistance is a little poor. Due to the existence of polar group, the electric insulating property decreases; phenol formaldehyde resin has certain toxicity.

Phenol formaldehyde resin possesses good wear resistance. Under the insufficient lubrication condition (oil deficiency or oil supply with short-time interruption), it still has high wear-resistance. Under good lubrication condition (especially water lubricating), the effect is more ideal. Its wear increases with the increase of load, while the friction coefficient decreases with the increase of load (0.15—0.05). The kinetic friction coefficient is 0.22 when phenol formaldehyde resin rubs with steel.

Phenol formaldehyde resin used as friction parts material is well known because it can bear high load; it can work well under the PV value of 0.53MPa·m/s. The PV value of reinforced phenol formaldehyde can reach 1—1.5MPa·m/s (unlubricated), 7.5—12MPa·m/s (oil lubricating), and even 30—100MPa·m/s (water lubricating).

Phenol formaldehyde resin used as friction piece has three shortcomings: ① The creeping will occur under low speed and small displacement condition. ② Seizure is easy to happen after the heat is generated due to friction. ③ When iron chip and other fluffs come into friction interface, they will be grinded to globularity, and they can easily damage the friction surface.

1.5.6 Epoxide Resin

The high-molecular compounds containing epoxy group of ($\begin{smallmatrix} -CH-CH- \\ \diagdown \diagup \\ O \end{smallmatrix}$) are called by a general name, epoxide (EP) resin. It is a high polymer compound obtained by the fasculation reaction of compound, which can generate epoxy groups during reaction process, and multi-hydroxy compound. This kind of polycondensate is a linear macromolecule and possesses thermal plasticity; however, as its molecular chains have many active groups, under the action of various solidifying agents (generally amine type or anhydride type), —CH in the macromolecule and epoxy groups at the two ends participate in the cross linking reaction; the linear structure turns into bulk structure; after heated to mould, the thermal plasticity becomes thermosetting property.

Epoxide resin is easy to solidify without water and other byproducts released. After solidification, the mechanical strength is very high; the dimension is stable; it can resist acid and alkali; its electric insulativity is also good. Epoxide resin has rather high wear-resistance; meanwhile it can greatly reduce the wear of counterface. It has rather low friction coefficient, which is only 1/7 that of cast iron; the static and kinetic friction coefficients are close, which can effectively prevent creeping. Under poor lubrication and high load condition, epoxide resin can also work normally; there is no impact and bounce when the motion is started; it can realize self-lubricating when oil is absent; it has high load-carrying capacity and good damping performance; meanwhile, the application is simple.

1.6   Conclusion

Different kinds of solid lubricants can play lubricating role in different conditions. Among them, sulfide lubricants such as FeS, MoS_2, WS_2, and ZnS are commonly applied. Sulfide used as solid lubricant has a hexagonal layered structure and low shearing force. It can easily slip along the close-packed planes and exhibits good solid lubricating properties. Especially, solid lubrication sulfide films or coatings prepared by newly technique show excellent anti-friction and wear resistance, which will be introduced in next chapters detailedly.

References

1. Shi M S. Solid Lubrication Materials [M]. Beijing: Chemical Industry Press, 2000.

2. Chen F M. Low-temperature electrolyzation sulfurizing of hydraulic parts [J]. Mechanical Engineer, 2005, (7): 147–149.

3. Du M Y, Xiang N, Zhu Z X, et al. Tribological characteristics of the surface layer of grey cast iron modified by ion nitriding & ion sulfurizing [J]. Heat Treatment of Metals, 2005, 30(5): 12–15.

4. Cheng J H, Ge P Q, Liu M. Experimental study on the friction and wear behavior of sulphurized 45 steel under dry friction condition [J]. Lubrication Engineering, 2002, (5): 35–36.

5. Wang H D, Zhuang D M, Wang K L, et al. Study on tribological properties of iron sulfide coating prepared by sol-gel method [J]. Journal of Materials Science Letters, 2003, 22(20): 1603–1606.

6. Zhu M H, Zhou H D, Chen J M, et al. A comparative study on radial and tangential fretting damage of molybdenum disulfide bonded solid lubrication coating [J]. Tribology, 2002, 22(1): 14–18.

7. Yu X, Wang C B, Jiang G W, et al. Tribological mechanism and property of 9Cr18 friction pair at atmosphere and in vacuum [J]. Vacuum, 2004, 72: 461–466.

8. Li Y L, Kin S K. Tribology and wear of MoS_2/Ti composite films [J]. Chinese Journal of Vacuum Science and Technology, 2005, 25(5): 378–380.

9. Su Y L, Kao W H. Tribological behaviour and wear mechanism of MoS_2-Cr coatings sliding against various counterbody [J]. Tribology International, 2003, 36: 11–23.

10. Zheng C B, Xu H M, Yang H, et al. Overview of film deposition technology of ion beam sputtering [J]. Laboratory Science, 2007, (4): 153–156.

11. Liu D X, Tang B, Chen H, et al. MoS_2 composite films on Ti alloy prepared by ion-beam-enhanced deposition [J]. The Chinese Journal of Nonferrous Metals, 2001, 11(3): 454–460.

12. Nainaparampil J J, Phanil A R, Krzanowski J E, et al. Pulsed laser ablated MoS_2-Al films: Friction and wear in humid conditions [J]. Surface and Coatings Technology, 2004, 187: 326–335.

13. Xu H Y, Zhou H D, Chen J M, et al. Properties of PA (polyamide) coating and PA/MoS_2 composite coatings by flame spraying [J]. Transactions of the Nonferrous Metals Society of China, 2004, 14: 67–71.

14. Smorygo O, Voronin S, Bertrand P, et al. Fabrication of thick molybdenum disulphide coatings by thermal-diffusion synthesis[J]. Tribology Letters, 2004, 17: 723–726.

15. Voronin S, Smorygo O, Bertrand P, et al. Thermal-diffusion synthesis of thick molybdenum disulphide coatings on steel substrates [J]. Surface and Coatings Technology, 2004, 180/181: 113–117.

16. Hu K H, Wo H Z, Han X Z, et al. Survey of preparing nanoscaled molybdenum disulfide [J]. Modern Chemical Industry, 2003, 23(8): 14–21.

17. Gao B, Zhao P. Syntheses and applications of inorganic fullerene-like molybdenum disulfide nanoparticles [J]. China Molybdenum Industry, 2007, 31(1): 49–53.

18. Shan J H, Ning G L, Lin Y. The applications and novel methods of preparation of molybdenum disulfide [J]. Science & Technology in Chemical Industry, 2002, 10(3): 42–45.

19. Martin-Litas I, Vinatier P, Levasseur A, et al. Characterisation of r.f.sputtered tungsten disulfide and oxysulfide thin films [J]. Thin Solid Films, 2002, 416: 1–9.

20. Devadasan J J, Sanjeeviraja C, Jayachandran M. Electrodeposition of p-WS$_2$ thin film and characterization [J]. Journal of Crystal Growth, 2001, 226: 67–72.
21. Zhang L L. Preparation and tribological properties of WS$_2$ nanorod as oil additive [D]. Hangzhou: Zhejiang University, 2006.
22. Feldman Y, Zka A, Popovitz-Biro R, et al. New reactor for production of tungsten disulfide hollow onion-like (inorganic fullerene-like) nanoparticles [J]. Solid State Science, 2000, 2: 663–672.
23. Tang G H. Present situation and prospect of synthesis and applications of ZnS nano-particle [J]. Hebei Chemical Industry, 2007, 30(9): 17–19.
24. Wang P F, Yuan Y, Liu H, et al. Research progress on the preparation of ZnS nano-particles [J]. Chemical World, 2003, (8): 441–444.
25. Li Y F, Lan Y Z. Preparation and application of nanocrystalline zinc sulfide [J]. Hydrometallurgy of China, 2007, 26(3): 123–127.
26. Murali K R, Dhanemozhi A C, John R. Brush plated ZnS films and their properties [J]. Journal of Alloys and Compounds, 2007, 464(1–2): 383–386.
27. Meng F, Zhang G Y, He J P, et al. ZnS thin films prepared by chemical bath deposition [J]. Chinese Journal of Power Sources, 2007, 31(12): 998–1000.
28. Jiu Z X. Influence of the laser pulse energy density on ZnS thin film [J]. Journal of Wuhan Polytechnic University, 2007, 26(2): 104–106.
29. Wang B Y, Zhang R G, Wan D Y, et al. Electroluminescent ZnS-based thin films and their preparation [J]. Materials Review, 2003, 17(11): 33–35.

Chapter 2
Solid Lubrication FeS Film Prepared by Ion Sulfuration

Many studies indicated that sulfide layer on the steel surface or application of sulfur-containing additives in lubricating oil can obviously improve the wear-resistance of workpieces, especially can effectively prevent their scuffing or seizure. This demonstrates that FeS is an effective solid lubricant. The electrolytic sulfuration has been used for years to produce sulfide layers, which greatly improved the anti-scuffing behaviour of friction pairs. However, the pollution problem caused by CN^- limited its propagation.

In the last 20 years, a new low temperature ion sulfuration process has quickly been developed, which can overcome the drawbacks of the electrolytic sulfuration process for environment protection. The sulfur atoms were ionized by glow discharge in the vacuum chamber and the sulfur ions diffused through the crystal defects and grain boundaries of the steel surface and reacted with iron ions to form the FeS film.

FeS is an excellent solid lubricant, with high melting point and low shearing strength. It possesses a hexagonal close-packed structure, with good properties of anti-friction and wear-resistance. The low friction coefficient is associated closely with its lamellar structure and weak interplanar bonding. By now, the solid lubrication FeS films have been studied extensively. However, the industrial community still lacks the understanding of the effective anti-friction and wear-resistance properties of FeS film. Some basic knowledge of FeS film prepared by ion sulfuration is described in this chapter.

2.1 The Microstructure of Solid FeS

The commercial FeS bulk was used in the research; the mass percent of FeS was not less than 75%. FeS particles with 1mm in diameter were obtained by grinding, while the FeS fine powders in the range of 1μm and 10μm were obtained by ball milling [1]. Figure 2.1 shows the macro-morphologies of different shapes of FeS.

2.1.1 Surface Morphologies of Solid FeS

The surface morphologies of different shapes of FeS are shown in Figure 2.2. It can be found from Figure 2.2 (a) that the surface of FeS bulk is relatively flat but a little rough, and there are some irregular polygonal particles on it, whose granularity is within 10μm, and some micro-cracks are also present. Figure 2.2 (b) demonstrates, the

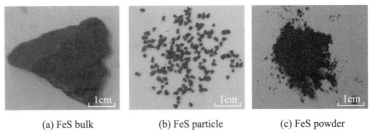

(a) FeS bulk   (b) FeS particle   (c) FeS powder

Fig.2.1   The macro-morphologies of different shapes of FeS

surface of FeS particle is comparatively smooth, it is a bit rugged, a few micro-pores can be seen, and the structure is looser than that of FeS bulk. Figure 2.2 (c) shows that the FeS powder is the loosest among the three shapes. They are usually piled up from a lot of small particles; some of them are inclined to form flocculent aggregation.

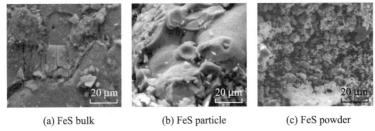

(a) FeS bulk   (b) FeS particle   (c) FeS powder

Fig.2.2   Surface SEM morphologies of different shapes of solid FeS

2.1.2   Phase Structures of Solid FeS

The XRD patterns of different shapes of solid FeS are shown in Figure 2.3. The principal phase is FeS with strong peaks; some weak peaks come from impurities. The width of diffraction peaks broadens gradually due to the reduction of grain size and lattice strain, which indicates that heavy deformation has been introduced into the grains of FeS particle and FeS powder during the grinding and ball milling process. The three shapes of FeS all have a hexagonal crystal structure, the lattice constants are $a = 0.597$nm, $c = 1.176$nm. That means, the lattice doesn't vary after grinding and milling.

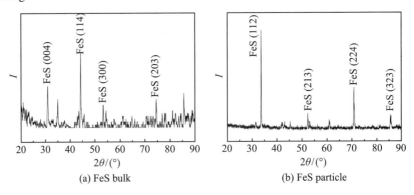

(a) FeS bulk   (b) FeS particle

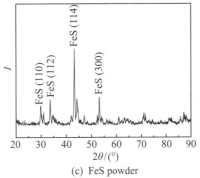

(c) FeS powder

Fig.2.3   XRD patterns of different shapes of solid FeS

2.1.3   TEM Morphologies of Solid FeS

The TEM morphologies of different FeS are shown in Figure 2.4. They display irregular laminal structure and disordered distribution with the diameter of grains in the range of 0.1μm to 3μm.

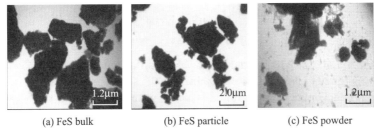

(a) FeS bulk   (b) FeS particle   (c) FeS powder

Fig.2.4   TEM photographs of different shapes of solid FeS

2.1.4   Analysis of Electron Diffraction

Figure 2.5 shows the photographs of electron diffraction of solid FeS. It can be seen that they are composed of many single and multi-crystals. The sulfur atoms and iron atoms constitute a hexagonal reticular structure with the lattice constant $a = 0.597$nm and $c = 1.176$nm. The close-packed planes are joined by weak van der waals force, so the FeS structure is easy to slip along the plane (001) by the shearing force.

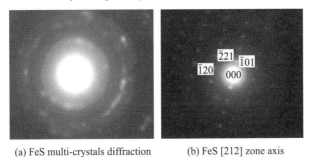

(a) FeS multi-crystals diffraction   (b) FeS [212] zone axis

Fig.2.5   Electron diffraction patterns of FeS

2.2   The Formation of Iron Sulfuration Layer

The published papers about low temperature ion sulfuration technique mostly focused on the process exploration and performance test of sulfide layers so far. The researches on the microstructures and forming process of sulfurized layers are still insufficient. Obviously, it is significant to make clear the formation and growth processes of the sulfide layers, which is necessary for controlling the process parameters and ensuring the quality of the lubricating layers. In this section, the change of microstructures and compositions of the sulfide layers on 1045 and 52100 steels with different sulfurizing times as well as the formation and growth processes were studied.

2.2.1   Experimental Methods

Experimental materials were AISI 1045 steel treated by quenching and high-temperature tempering, with a hardness of HRC 26—30, as well as AISI 52100 steel treated by quenching and low-temperature tempering, with a hardness of HRC 57—62. The surfaces of specimens were polished carefully before ion sulfurization. The sulfurizing temperatures were 160—200°C. The holding times were 15min, 30min, 60min, 90min, and 120min. SEM equipped with EDX was used to analyze the morphology and composition of the sulfide layer. XRD was employed to study its structure.

2.2.2   Surface Morphologies of Sulfuration Layers

Figures 2.6 and 2.7 show the morphologies of sulfurized layers on 1045 steel with sulfurizing time of 15min, 60min, 120min, as well as on 52100 steels with sulfurizing time of 30min, 90min, 120min. A lot of cavities were present on the surface of sulfide layer, which were especially big and deep at the initial forming stage; their maximum diameter could reach 5μm, with the shape of "volcanic crater" as shown in Figure 2.8. There were obvious sulfide depositions around the "volcanic crater". Some Small and shallow pits also appeared on the layer surface, as shown in Figure 2.9. With the increase of sulfurizing time, the cavity diameter became enlarged. Finally, the layer surface showed a character of many honeycomb-like small and loose holes, as shown in Figure 2.10. The cavities on the surfaces of sulfide layers on 1045 steel at 120min and 52100 steel at 90min became smaller and shallower, and "volcanic crater" shape disappeared. Figure 2.11 shows the surface indentation morphologies of the sulfide layers

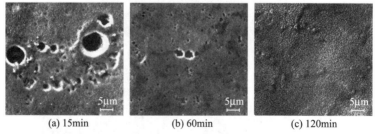

(a) 15min   (b) 60min   (c) 120min

Fig.2.6   Morphologies of the sulfurized layers on 1045 steels

on 1045 steel and 52100 steel at a load of 200g as well as the scratching micrograph of sulfide layer on 52100 steel measured by a micro-hardness tester. The white spots on the surface of the spalling area on 52100 steel were chromium carbides. The composition analysis for the surfaces of spalling areas and original layers were listed in Table 2.1. The sulfur content on the spalling area decreased obviously, and almost no sulfur could be detected, that indicated, no transition zone existed between the sulfide layer and substrate.

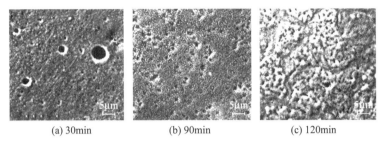

| (a) 30min | (b) 90min | (c) 120min |

Fig.2.7   Morphologies of the sulfurized layers on 52100 steels

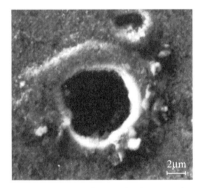

Fig.2.8   "Volcanic crater" shape of the cavities on the layer surface

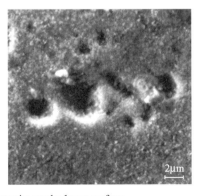

Fig.2.9   Small and shallow pits on the layer surface

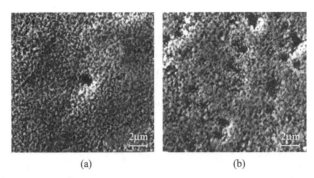

(a)   (b)

Fig.2.10   The cavities on the surface of sulfide layers on (a) 1045 steel at 120min and (b) 52100 steel at 90min became smaller and shallower, and "volcanic crater" shape disappeared

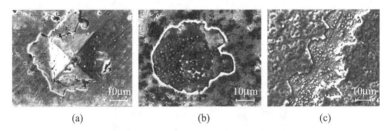

(a)   (b)   (c)

Fig.2.11   Surface indentation morphologies of the sulfide layers on (a) 1045 steel and (b) 52100 steel at a load of 200g as well as the scratching micrograph of sulfide layer on (c) 52100 steel measured by a micro-hardness tester

Table 2.1   Composition analysis for the surfaces of spalled layers and original layers

Testpieces	Atom%	Fe	S	Cr
1045 steel	Sulfurized layer surface	86.64	13.36	—
	Spalling area	99.36	0.64	—
52100 steel	Sulfurized layer surface	69.87	28.35	1.76
	Spalling area	96.64	1.37	1.98

2.2.3   Composition on the Sulfurized Steel Surface

Figure 2.12 shows the sulfur contents on the sulfurized 1045 and 52100 steels surface with different sulfurizing time. With the increase of time, the sulfur content is going up. The sulfur content on the sulfurized 52100 steel surface was higher than that on the 1045 steel after same sulfurizing time. Figure 2.13 shows the atomic concentration ratio Fe/S on sulfurized 1045 and 52100 steels surface with different sulfurizing time. With the increase of time, the concentration ratio Fe/S was decreased gradually. Owing to the ratio Fe/S on the sulfurized 52100 steel surface has been approaching to 1 when sulfurizing time is over 90min, it can be considered that a complete sulfide layer has been formed on the substrate when the concentration ratio Fe/S was close to 1. In addition, the concentration ratio Fe/S on the sulfurized 52100 steel surface was less than that on 1045 steel.

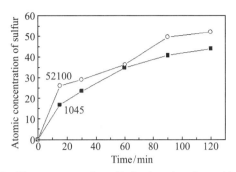

Fig.2.12   Variation of sulfur content on the sulfurized steel surface with sulfurizing time

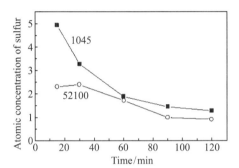

Fig.2.13   Variation of atomic concentration ratio Fe/S on the sulfurized steel surface with sulfurizing time

2.2.4   Phase Structure of Sulfide Layer at Different Sulfurizing Time

The phase structures of sulfide layer at different sulfurizing time are shown in Figure 2.14. The sulfurized layer was mainly composed of α-Fe, FeS and FeS$_2$. For the 1045 steel, the sulfide was mainly FeS at 15min. With the increase of time, the intensity of

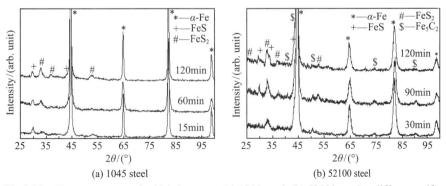

Fig.2.14   Phase structures of sulfide layers on (a) 1045 steel, (b) 52100 steel at different sulfurizing times

FeS$_2$ became stronger gradually. For 52100 steel, the sulfide was mainly FeS$_2$ in the beginning, and then the intensity of FeS enhanced gradually with increasing time. A weak phase of Fe$_5$C$_2$ was also detected, but it became reduced at longer sulfurizing time.

2.2.5 Formation Mechanism of Sulfurized Layer

The basic principle of ion sulfurizing is quite similar to that of ion nitriding, but they have difference at some points. The configuration of extra-nuclear electron of sulfur is 1s^{2}2s^{2}2p^{6}3s^{2}3p^4, it shows a stronger chemical activity, but its atomic radius ($r_S = 1.02$Å) is much bigger than that of nitrogen ($r_N = 0.75$Å); therefore, the diffusion of sulfur atoms into iron is very difficult; the interstitial solid solution almost can not be formed. At the initial stage of formation, it is mainly dependent on the adsorption and deposition of sulfides but not on the diffusion. FeS is generated by the reaction of sputtered iron ions and sulfur ions and then deposited on the substrate surface. In the process of sulfurizing, the sulfide is formed not only by deposition, but also by diffusion of sulfur into the substrate through lattice defects and grain boundaries. The serious lattice distortion can be occurred on the sample surface due to the ion bombardment and the dislocation density can be increased significantly. In addition, a plenty of pits with different size can be also generated, especially at the grain boundaries. Obviously, such a high degree of activation of the surface is beneficial to the formation of sulfides. The initially formed sulfide nuclei continue to grow around and meet other sulfide nuclei when they reach a certain size, which will lead to the appearance of cavities at the junctions. With the increase of sulfurizing time, sulfides become expanded and thickened gradually and make the cavities smaller and smaller, finally, a character of honeycomb-like small and loose holes is formed. With the increase of sulfide thickness, the reaction of sulfur and iron gradually becomes weakened. Some researchers have studied the relationship between sulfurizing time and thickness of sulfide layer, as shown in Figure 2.15. It indicates that there is a parabola relationship between sulfurizing time and layer thickness. The thickness increase becomes not obvious after a certain sulfurizing time. Therefore, it is unnecessary to pursue an excessive thickness of the FeS layer.

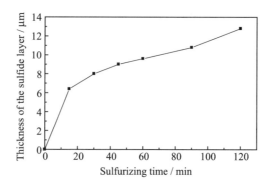

Fig.2.15   Relationship between sulfurizing time and thickness of sulfide layer

The XRD results indicate that FeS is firstly formed on 1045 steel, while FeS_2 on 52100 steel. With the increase of time, FeS_2 on the former and FeS on the latter are increased gradually. Although the curing rate of Cr is about 1—2 orders of magnitude lower than that of Fe, a variety of sulfides can be also formed by reaction of Cr and S, such as Cr_2S_3, CrS, and Cr_3S_4. Meanwhile, 52100 steel contains more carbides, and sulfur atoms might replace carbon atoms to yield some active metal atoms, that is also beneficial to the adsorption and deposition of sulfides. Therefore, the sulfur content on the sulfurized 52100 steel surface is higher and FeS_2 is firstly formed at the initial stage. While the sulfur content on 1045 steel surface is lower and FeS is formed preferentially.

2.3   Characterization of Ion Sulfurized Layer

2.3.1   Characterization of Sulfurized Layer on 1045 and 52100 Steels

2.3.1.1   Microstructures

Figures 2.16 and 2.17 are the surface morphologies of 1045 steel sulfuration layer observed under AFM and SEM. The FeS films were all loose and porous, which were formed with nano spherical grains. The granularity of all grains was less than 100nm.

The surface morphologies of sulfide layers with 4μm and 12μm thickness on 52100 steel are shown in Figure 2.18. The layers were also porous and loose. The grains of 4-μm layer looked rather shallow, while those of 12-μm layer looked much deeper. It indicated that the microporosities were formed gradually with the increase of layer thickness. The high magnification morphologies showed that the layer was composed of a great deal flocculent particles, which had nothing to do with the thickness. Figure 2.19 shows the cross-section morphologies of the sulfide layers on 1045 and 52100 steels. It was clear that no transition zone was found between the layer and substrate. Figure 2.20 is the distribution map of S element of sulfide layer on 52100 steel. The distribution maps of Fe, S, and Cr elements of sulfide layer on 52100 steel are shown in Figure 2.21.

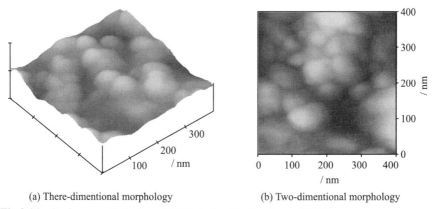

(a) There-dimensional morphology   (b) Two-dimentional morphology

Fig.2.16   Surface morphologies of 1045 steel sulfuration layer (AFM)

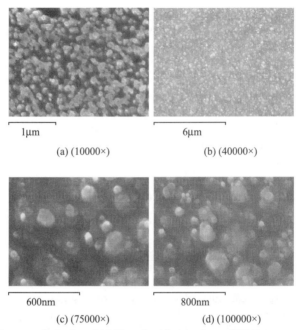

(a) (10000×)   (b) (40000×)

(c) (75000×)   (d) (100000×)

Fig.2.17   Surface morphologies of 1045 steel sulfuration layer (SEM)

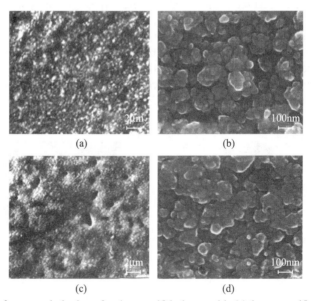

(a)   (b)

(c)   (d)

Fig.2.18   Surface morphologies of a 4-μm sulfide layer with (a) low magnification and (b) high magnification, and a 12-μm layer with (c) low magnification and (d) high magnification on 52100 steel

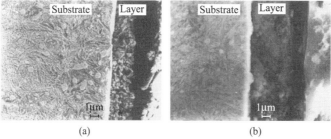

(a)   (b)

Fig.2.19 Cross-section morphologies of sulfide layers on (a) 1045 steel and (b) 52100 steel

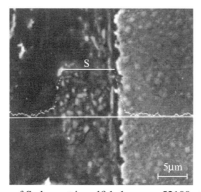

Fig.2.20 Distribution map of S element in sulfide layer on 52100 steel

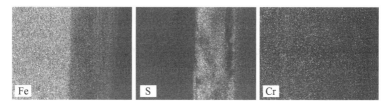

Fig.2.21 Distribution maps of Fe, S and Cr elements in sulfide layer on 52100 steel

2.3.1.2 Phase structures of the sulfide layer

Two XRD analysis techniques, "the conventional θ-2θ scanning or symmetric Bragg-diffracting geometry" and "the grazing-incident X-ray and asymmetric-Bragg diffraction (GXRD)" were employed. The basic structure of the sulfide layer was first identified by the conventional θ-2θ scanning technique, where the incident angle α was equal to the diffraction angle θ, as shown in Figure 2.22. Provided that the intensities of incident and diffraction beams were I_0 and I ($I < I_0$), respectively, then $I = I_0 \, e^{-\mu(2d/\sin\theta)}$, where d is the penetration depth, μ is the X-ray linear absorption coefficient. Because $\mu_{Fe} \geqslant \mu_S$ and Fe is the main element in the layer, the absorption coefficient of Fe to Cu Kα ray is regarded as that of the sulfide layer. If the depth of $I/I_0 = 13\%$ was regarded as the penetration depth of X-ray, then the penetration depth of Cu Kα X-ray was approximately 4µm. In other words, this technique identified the microstructure within a 4-µm depth of the sulfide layer from the top surface.

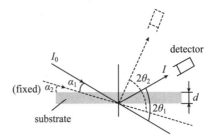

Figure 2.22 Schematic representation of X-ray analysis, solid line represents the symmetric Bragg-diffraction geometry ($\alpha = \theta$), and dash line represents the GIABD geometry ($\alpha \neq \theta$)

GXRD is an effective technique for determining the crystalline phases on surface and structural depth profiles of the graded layers. The thin-film specimen was fixed at a grazing angle with respect to the incident X-ray beam. The detector was scanning along the 2θ-circle in the vertical plane of a diffractometer to record the asymmetrically Bragg diffracted X-rays from crystal planes inclined to the specimen surface. A fixed incident angel α of a few degrees was used so that the path in the film can be lengthened or the penetration depth of the X-rays can be decreased by an order of magnitude. The GXRD technique has successfully been used for characterizing single- and multi-layer thin films. In this study, the low incident angle $\alpha = 1°$, $1.5°$, $2°$, $2.5°$, $3°$, $4°$, $5°$, $7°$, $9°$, $10°$, $11°$ were selected. The surface roughness of the specimen was lower than $R_a = 0.04$mm, and its planarity was lower than 1μm so that the influence of surface topography on the analysis results could be ignored. Corresponding to $\alpha = 1°$, $1.5°$, $2°$, $2.5°$, $3°$, $4°$, $5°$, $7°$, $9°$, $10°$, $11°$, the penetration depth $d = 0.14\mu$m, 0.21μm, 0.28μm, 0.35μm, 0.42μm, 0.56μm, 0.70μm, 0.98μm, 1.25μm, 1.39μm, 1.53μm, respectively.

GXRD analysis was conducted on a high-resolution X-ray diffraction diffractometer of model D8 Discover, its core component is a high precision two-circle goniometer, with two separate stepper motor controlling the movement of θ and 2θ. The goniometer and its accessories are controlled by a microprocessor control unit, the operation of which is controlled by a manual control panel or computer. The collimation of the diffractometer is also controlled by computer. The parameters are as follows.

Measuring circle diameter: 500mm and 600mm;

Resolution of θ-circle: $0.0001°$;

Resolution of 2θ-circle: $0.0002°$;

Accuracy of 2θ-circle: $< +0.0002°$.

The phase constitutions of the sulfide layers on 52100 and 1045 steels are shown in Figure 2.23. The high diffraction intensity of α-Fe phase indicated that the main phase in the layer was α-Fe. The sulfide was composed of FeS and FeS_2. With the increase of layer thickness, FeS_2 with an orthorhombic structure was increased. Figure 2.24 shows the analysis results of GXRD technique. That indicated, only FeS_2 and α-Fe appeared on the top surface. With the increase of penetration depth, FeS was growing up.

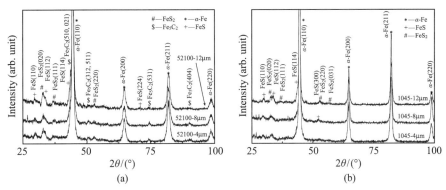

Fig.2.23   Phase constitutions of the sulfide layers with 4-μm, 8-μm, and 12-μm thickness on (a) 52100 and (b) 1045 steels

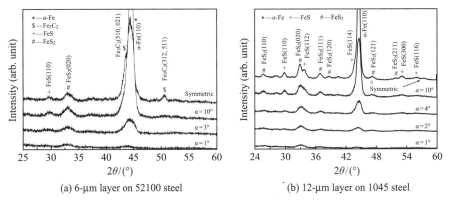

Fig.2.24   Variation of phase constitutions with increase of depth below top surface of the sulfide layer

(α is the X-ray incident angle; High α values represent deeper X-ray penetration)

2.3.1.3   Depth analysis of a 3μm sulfide layer on 52100 steel

Depth chromatography technique is dependent on that different incident angles reflect different depth information. The XRD pattern of a particular thin layer can be obtained through calculation, including all the information, such as intensity, peak position and peak shape. Therefore, more detailed qualitative and quantitative researches can be carried out.

Depth chromatography technique was used to analyse the structure and composition distribution of sulfide layer with a thickness of 3μm on 52100 steel. The conventional θ-2θ symmetrical pattern of sulfide layer is shown in Figure 2.25. Owing to the information depth of symmetrical incident was 4μm, which reflected only the average structure information at 4μm below the surface, the distribution of phase structures in depth from the surface could not be determined. Therefore, the structure distribution in depth was obtained by changing the incident angles. Figure 2.26 shows the measured XRD pattern of sulfide layer on 52100 steel with different incident angles, they are 1°, 1.5°, 2°, 2.5°, 3°, 4°, 5°, 7°, 9° and 11° from top to bottom respectively. The intensity of FeS_2 was higher at 1° and became weakened with the increase of incident angle, but the intensity of FeS was enhanced. When the incident angles further increased, the intensity of both sulfides decreased and the pattern reflected more deep-seated structure information. Therefore, as the incident angle became larger, the proportion of the surface information was reduced and the effect of inner phase on the pattern was enhanced. Figure 2.27 shows the calculated XRD pattern of sulfide layer on 52100 steel with different incident angles. Each pattern directly reflected the structure information of each depth, therefore, the variation of content of each phase with depth could

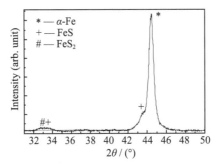

Fig.2.25   Conventional θ-2θ symmetrical pattern of sulfide layer on 52100 steel

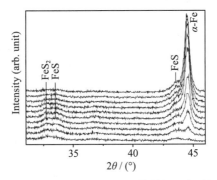

Fig.2.26   Measured XRD pattern of sulfide layer on 52100 steel with different incident angles ($\alpha = 1°, 1.5°, 2°, 2.5°, 3°, 4°, 5°, 7°, 9°, 11°$)

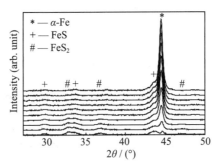

Fig.2.27   Calculated XRD pattern of sulfide layer on 52100 steel with different incident angles ($\alpha = 1°, 1.5°, 2°, 2.5°, 3°, 4°, 5°, 7°, 9°, 11°$)

be observed directly. FeS_2 was the main phase on the surface of layer and the content of α-Fe was not high. A certain content of FeS was also found. With the depth increasing, the content of α-Fe increased. Figure 2.28 shows the cross-section morphology of sulfide layer. Figure 2.29 shows the distributions of Fe, S elements from 52100 steel substrate to sulfide layer. The content of sulfur was gradually reduced from surface to inner, while iron showed a contrary trend, which was almost identical to the information obtained from the calculated pattern of Figure 2.27.

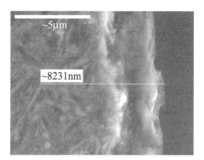

Fig.2.28   Cross-section morphology of a 4μm thick sulfide layer on 52100 steel

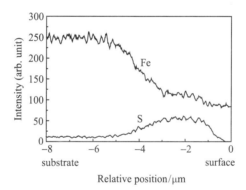

Fig.2.29   Distributions of Fe, S elements on the cross-section of 52100 steel and sulfide layer

2.3.1.4   Residual stress of the sulfide layer on 1045 steel

The residual stress of the sulfide layer on 1045 steel was measured by X-ray stressometer. Figure 2.30 shows the relation between 2θ and $\sin^2 \Psi$. The scanning angles ranged between $155°$ and $158°$. The inclination angles Ψ were $0°$, $25°$, $35°$, $45°$, respectively. The stress value measured was -150MPa, which indicated that the compressive residual stress was present on the surface of sulfide layer. As well-known, the residual stress can obviously affect the film's quality and performance. The residual compressive stress can relax the stress concentration and improve the anti-fatigue performance of the film and bonding strength between the film and substrate. However, the excessive compressive stress can also induce the delamination and peeling. In this experiment, the residual compressive stress of the FeS film was relatively moderate, therefore, the cracks and delamination were not found.

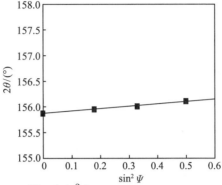

Fig.2.30   Relation between 2θ and $\sin^2 \psi$

2.3.1.5   Mechanical properties of sulfide layer on 1045 steel

The curves of depth vs. load at five different positions measured by a nano-indentation tester are shown in Figure 2.31. The five curves were nearly the same, which indicated that the distribution of the sulfide layer along depth was quite homogeneous.

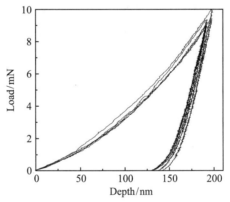

Fig.2.31   The curves of depth vs. load at five different positions

Table 2.2 shows the nanohardness and elastic modulus of 1045 steel and FeS film. The hardnesses of 1045 steel and FeS film were 10.51GPa, 2.84GPa, respectively. The elastic modulus of FeS film was 52.97GPa.

Table 2.2   Nanohardness and elastic modulus of 1045 steel and FeS film

Samples	1045 steel	FeS film
Nanohardness/GPa	10.51	2.84
Elastic modulus/GPa	247.76	52.97

2.3.2   Characterization of Sulfurized Layer on Four Kinds of Steels

In order to extend the application of ion sulfuration technique and examine its universality in more kinds of steel, four representative steels, namely high-speed steel, die steel, stainless steel and 1045 steel were selected to be treated by low temperature ion sulfuration [2].

The four steels for experiments were 1045 steel, M2 steel, and 420 steel, L6 steel. The chemical compositions and hardnesses are shown in Table 2.3. The surface roughness R_a was 0.04μm after polishing.

Table 2.3 Chemical composition and hardness of four steels

Steel	Hardness (HRC)	Chemical composition/%							
		C	Mn	Si	Cr	W	Mo	V	Ni
M2	65	0.80—1.00	< 0.35	< 0.30	3.50—4.50	6.0—7.0	4.0—6.0	1.80—2.30	
420	40	0.16—0.24			12.0—14.0				
L6	58	0.50—0.60	0.50—0.80	< 0.35	0.50—0.80		0.15—0.30		1.40—1.80
1045	55	0.42—0.50	0.50—0.80	0.17—0.37					

Figure 2.32 shows the cross-section morphologies of sulfide layers on four steels. The sulfide layer was displayed as a continuous black narrow band, their thicknesses ranged between 6μm and 10μm, no transition zone was found.

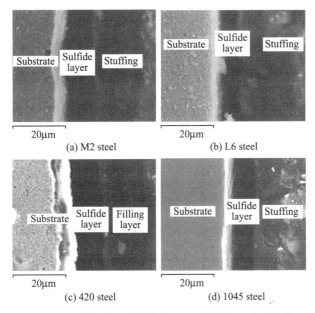

(a) M2 steel   (b) L6 steel

(c) 420 steel   (d) 1045 steel

Fig.2.32 Cross-section morphologies of FeS films on (a) M2 steel, (b)L6 steel, (c) 420 steel and (d) 1045 steel

Figure 2.33 shows the X-ray diffraction patterns of sulfide layers of four steels. Both FeS and FeS_2 were produced, and it should be noticed that owing to more contents of tungsten and molybdenum were contained in high speed steel, the solid lubrication phases of WS_2, MoS_2 were also formed in sulfide layer, which can further improve the solid lubrication effect.

Figure 2.34 shows the bonding strengths between sulfide layer and substrate for four steels. The peak suddenly occurred means that the bonding was destroyed at a certain load, and this load was regarded as the bonding strength. The bonding strength did not vary with the steel kind, which kept a higher value between 43.8N and 46.6N.

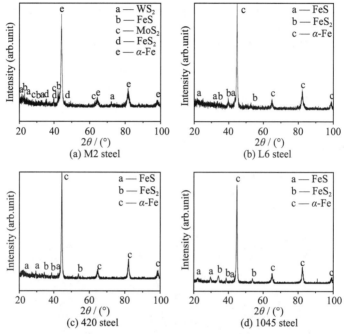

Fig.2.33   Phase structure of the sulfide layers by XRD

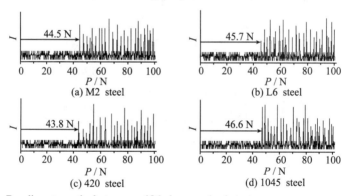

Fig.2.34   Bonding strengths between sulfide layer and substrate

2.4   Tribological Properties of Sulfurized Layers

2.4.1   Tribological Properties of Sulfurized Layers on 1045 and 52100 Steels

By now, low temperature ion sulfation technique has been applied for the modification of many machine parts, such as cutting tools, cylinder liners, piston rings, gears, bearings etc. and their service life has been extended significantly. However, the friction and wear mechanisms of sulfide layer and the tribological behaviors of sulfide layer have not been investigated deeply.

In this section, sulfide layers were made on 1045 and 52100 steel surfaces by means of low-temperature ion sulfurization technique. Their friction and wear behaviors under dry, liquid paraffin lubrication, and engine-oil lubrication conditions were investigated. The wear mechanisms of sulfide layer were approached in detail as well.

2.4.1.1   Experimental methods

Friction and wear tests were conducted on a SRV reciprocating tester under dry and liquid paraffin lubrication conditions. The upper samples were 52100 steel balls, 10mm in diameter with a hardness of HV 770. The lower samples were sulfurized and unsulfurized discs, with 24mm in diameter and 5mm in thickness. Its surface roughness was $R_a = 0.11$mm before sulfuration.

Experimental parameters under dry conditions: vibration amplitude was 1mm, frequencies were 15Hz and 30Hz, loads were 5N, 10N, 20N and 50N. In order to investigate the variation of sulfide layer during wear process, the test was stopped at different times at 50N, and the wear debris was collected.

Experimental parameters under paraffin oil lubrication conditions: vibration amplitude was 1mm, frequency was 15Hz, and load was 80N. The dynamic viscosity of paraffin oil is 14.7cSt at 40°C and its viscosity index is 87.

The widths of wear scars were measured by optical microscope. Morphologies and compositions of wear scars and wear debris were analysed by SEM and EDX.

The anti-scuffing performance was examined on a ball-on-disc tester. The upper specimen was a 52100 steel ball with 12mm in diameter and hardness of HV 770. The lower samples were sulfurized and unsulfurized 1045 steel discs with 60mm in diameter and 5mm in thickness. Its average surface roughness R_a before sulfuration was 0.8μm. The lubricant was engine oil with a dynamic viscosity of 37—43cSt at 50°C and the oil supplying rate was 1.9mL/min. After a running-in period for 1min, a load of 12.8kg was applied for 2min. The next loading increment was 5.8kg per 2min until scuffing occurred. Scuffing was indicated by a sudden increase of friction force with great vibration. The load at this time was regarded as the anti-scuffing load. P-v diagrams were plotted to link the anti-scuffing loads P at different sliding speeds v.

Optical microscopy was used to measure the width of wear scar of disc and ball. XPS and AES were employed to analyse the composition and valence state of compounds in boundary lubrication film.

2.4.1.2   Effect of preservation mode on the tribological properties of sulfide layers

Friction and wear tests were conducted on a SRV reciprocating tester. The variation of friction coefficient μ with time can be shown by a scheme in Figure 2.35. At initial stage of test, the coefficient μ kept a low and steady state. After a period of test, sulfide layer was worn away and lost its effect gradually, and then metallic surfaces came into direct contact, which lead to the sharp rise and large vibration of friction coefficient. The friction coefficient μ_b at initial stage, the endurance life t_c (when $\mu = 0.3$) of the layer, and μ_t at final stage were considered as the indexes to evaluate the tribological behaviors of sulfide layer. However, the μ of plain surface rose to a fairly high level immediately as soon as the test began, i.e., $t_c = 0$.

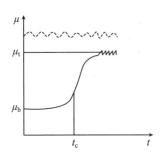

Fig.2.35 Variation of friction coefficient μ of sulfide layer (the solid line) and the plain surface (the dash line) with time t

Table 2.4 shows the tribological properties of sulfide layers tested after sulfuration and storage for 118 days. The friction coefficient of the sulfide layer stored in air was higher, the t_c was shorter and the wear scar width w was greater, and the tribological properties were worse than that tested just after sulfuration. On the contrary, the friction coefficient of the sulfide layer soaked in oil was the lowest, the t_c was the longest and the wear scar width w was the least. Therefore, the sulfide layer soaked in oil possessed the best tribological performance. The tribological properties indexes of the layers stored in the drier were close to that tested after sulfuration. It can be concluded that the sulfide layers should be preserved in oil, because when FeS contacted with oxygen and humid air, a series of possible reactions could occur as follows:

$$FeS + 2H_2O \rightarrow H_2S + Fe(OH)_2$$
$$4Fe(OH)_2 + O_2 + 2H_2O \rightarrow 4Fe(OH)_3$$
$$4FeS + 3O_2 + 6H_2O \rightarrow 4Fe(OH)_3 + 4S$$

or

$$4FeS + 7O_2 \rightarrow 2Fe_2O_3 + 4SO_2$$
$$4FeS + 2H_2O + 7O_2 \rightarrow 4FeO(OH) + 4SO_2$$

Table 2.4 Tribological properties of sulfide layers tested after sulfuration and storage for 118 days

Preservation mode	Tested after sulfuration	Placed in the drier	Placed in air	Soaked in oil
μ_b	0.12—0.15	0.10—0.13	0.14—0.16	0.08—0.09
t_c/s	8.3	8	6	15.8
μ_t	0.42—0.48	0.49—0.52	0.45—0.51	0.41—0.46
w/mm	1.19	1.17	1.20	1.14

Figure 2.36 shows the phase structures of sulfide layers tested after sulfuration and

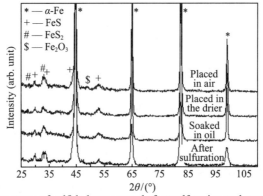

Fig.2.36 Phase structures of sulfide layers tested after sulfuration and storage for 118 days

storage for 118 days. Fe_2O_3 phase existed in the sulfide layer stored in air, while it was not present in the layers stored by other methods.

2.4.1.3 The tribological properties of sulfide layer under dry condition

1. The friction coefficient μ_b and μ_t of sulfide layer

Figure 2.37 shows the variation of friction coefficient with the thickness of sulfide layer on 1045 and 52100 steels. Figure 2.38 shows the variation of friction coefficient with load. At the initial stage, μ_b of the sulfurized surface was only about 30% that of the plain surface, μ_t was also slightly lower than that. This means, the sulfide layer possessed a remarkable friction-reducing effect. With the increase of layer thickness, μ_b was unchanged. As the load increasing, the friction coefficients of both sulfide layer and plain surface were decreased, but the decreasing trend was reduced, the μ_b kept steady to a certain value finally. μ_b of sulfide layer on 1045 and 52100 steels were about the same, which indicated that μ_b was independent of the substrate. At low loads like 5N, μ of the plain surface was higher, even more than 1.0, and fluctuated violently, that means, the adhesion became very severe at this moment. However, at high loads, μ became lower due to the alleviated adhesion. This phenomenon was probably related with the formation of oxide film due to friction heat, especially at high loads. Although μ of plain surface of 1045 steel was higher than that of 52100

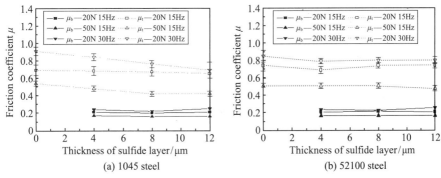

Fig.2.37 Variation of friction coefficient with the thickness of sulfide layer on (a) 1045 steel and (b) 52100 steel

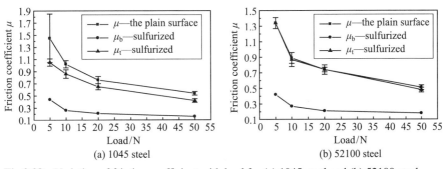

Fig.2.38 Variation of friction coefficient with load for (a) 1045 steel and (b) 52100 steel

steel, μ_t of sulfide layer on the former was less than that on the latter, as shown in Figure 2.39.

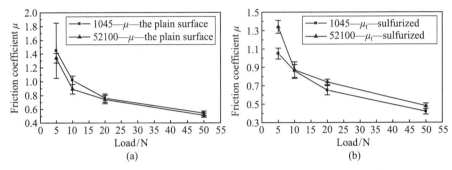

Fig.2.39   Variation of friction coefficient (a) μ of the plain surface and (b) μ_t of the sulfurized surface with load for 1045 and 52100 steels

2. Endurance life t_c of sulfide layer

Figure 2.40 shows the variation of endurance life t_c of sulfide layer with thickness. With the increase of layer thickness, the endurance life extended. The endurance life t_c of 12μm thick sulfide layer at different loads is shown in Figure 2.41. It can be seen that t_c was shortened with increase of load. The endurance life of sulfide layer on 52100 steel was longer than that on 1045 steel. But this phenomenon was not obvious

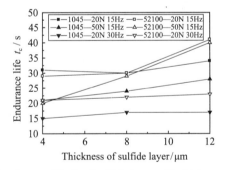

Fig.2.40   Variation of endurance life t_c of sulfide layer with thickness

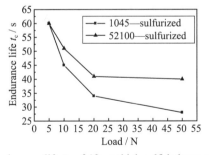

Fig.2.41   Variation of endurance life t_c of 12μm thick sulfide layer with load

at low load of about 5N. The difference between both cases became larger with the increase of load.

3. Wear scar width w of sulfide layer

The variation of width of wear scar with layer thickness is shown in Figure 2.42. The width of wear scar w decreased with the increase of layer thickness, indicating the improvement of wear-resistance. In addition, the width of wear scar of the sulfide layer on 52100 steel was smaller than that on 1045 steel. The widths of wear scars w of sulfide layers and plain surface at different loads are shown in Figure 2.43. It can be seen that the widths of wear scar of sulfide layers were lower than that of the plain surface. Figure 2.44 shows the variation of width of wear scar with time t at 50N. With regard to 1045 steel, the running-in period (when $\mathrm{d}^2 w/\mathrm{d}t^2 < 0$) of sulfurized surface was very short, and the steady stage (when $\mathrm{d}^2 w/\mathrm{d}t^2 = 0$) was reached soon after the test began. However, the steady stage was not reached yet after a relatively long running-in period for the plain surface. At the steady stage, wear rates $(\mathrm{d}w/\mathrm{d}t)$ of the sulfurized and plain surface were almost the same. With regard to 52100 steel, the wear rate of sulfurized surface was less than that of the plain surface at either running-in or steady stage. Therefore, the results mentioned above proved that the sulfide layer could promote the running-in process of surface, decrease wear rate, and improve wear-resistance of steels.

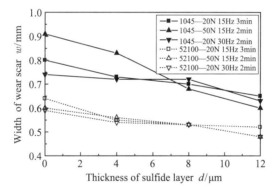

Fig.2.42   Variation of width of wear scar with layer thickness

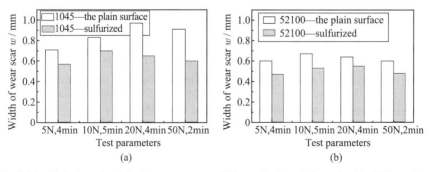

Fig.2.43   Variation of width of wear scar w at different loads and times for (a) 1045 steel and (b) 52100 steel

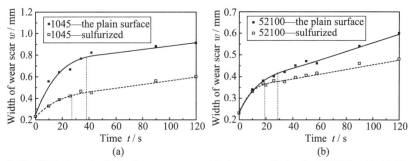

Fig.2.44   Variation of width of wear scar w with time t at the load of 50N for (a) 1045 steel and (b) 52100 steel, the dot lines represents the end of the running-in period

(When $t = 0$, w is the width of Hertz contact zone of 0.23mm)

4. Anti-scuffing behaviors of sulfide layer

Figure 2.45 shows the variation of anti-scuffing load P and friction coefficient μ with thickness of sulfide layer on 1045 steel under dry condition. The anti-scuffing load of sulfurized layer was obviously superior to that of unsulfurized sample. In addition, the friction coefficient of sulfurized layer was also lower than that of plain surface.

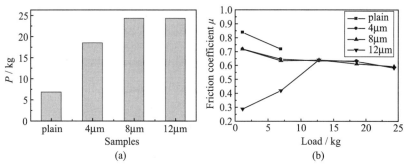

Fig.2.45   Variation of (a) anti-scuffing load P and (b) friction coefficient μ with thickness of sulfide layer on 1045 steel under dry condition

5. Morphology and composition of wear scars and wear debris

Figure 2.46 shows the morphologies of wear scar of plain surface and the sulfuration layer on 1045 steel as worn under dry condition. The composition of wear scar of sulfurized surface is shown in Figure 2.47. It can be seen clearly that the plain surface had been scuffed and worn severely while the wear scar of the sulfurized surface was still fine. A great amount of sulfur was found on the wear scar of the sulfurized surface and its distribution map as worn about 20s is shown in Figure 2.48. The large black film was just the adhesion film of sulfide. When the test was carried out about 2min, the scuffing of two samples occurred and sulfur almost did not exist on the worn surface of the sulfuration layer.

Figure 2.49 shows the wear scar morphologies of plain surface of 52100 steel as worn about 26s and 2min at 50N. Sulfurized surface of 52100 steel and the distribution maps of S element are shown in Figure 2.50. Compositions of the wear surface

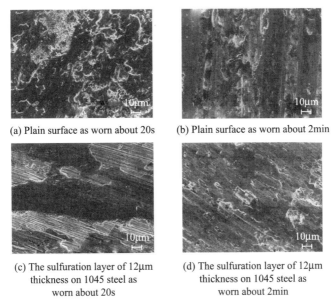

(a) Plain surface as worn about 20s   (b) Plain surface as worn about 2min

(c) The sulfuration layer of 12μm   (d) The sulfuration layer of 12μm
thickness on 1045 steel as   thickness on 1045 steel as
worn about 20s   worn about 2min

Fig.2.46   Morphologies of the wear scar of plain surface and sulfuration layer of 12μm thickness on 1045 steel as worn about 20s and 2min at 50N, 15Hz, 1mm under dry condition

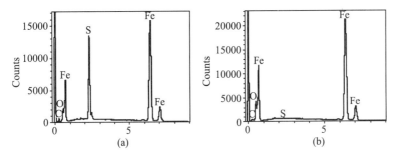

Fig.2.47   The composition of the wear scar of sulfurized surface as worn about (a) 20s and (b) 2min

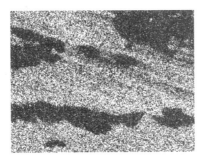

Fig.2.48   Distribution map of sulfur on the worn surface of sulfide layer as worn about 20s

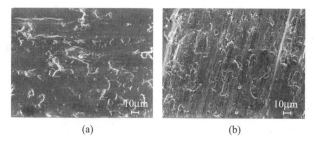

(a)   (b)

Fig.2.49   Morphologies of wear scars of the plain surface of 52100 steel at time of (a) 26s and (b) 2min under the load of 50N

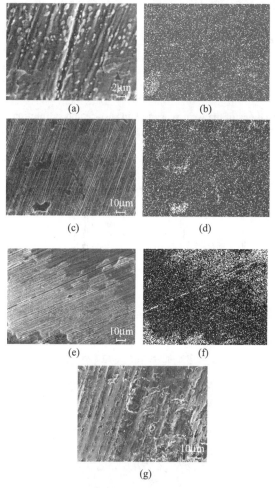

(g)

Fig.2.50   Wear scar morphologies of sulfurized surface of 52100 steel at the time of (a) 10s (c) 26s, (e) 42s and (g) 2min under the load of 50 N, and sulfur distribution maps (b), (d), (f) on the surface of (a), (c), (e), respectively.

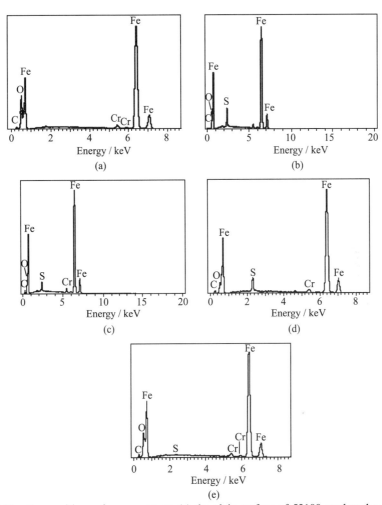

Fig.2.51 Compositions of wear scars on (a) the plain surface of 52100 steel at the time of 2min, and on the sulfurized surface at the time of (b) 10s, (c) 26s, (d) 42s and (e) 2min, corresponding to Figure 2.49 (b), Figure 2.50 (a), (c), (e) and (g), respectively

analyzed by EDX are shown in Figure 2.51. It can be seen that plastic deformation had taken place on the plain surface at 26s. When the test lasted for 2min, the wear became very severe and the content of O element increased obviously. However, only slight wear was found on the sulfurized surface, and its wear scar was fairly fine at initial stage of test. The wear surface was covered with sulfide film. With the increase of time, the sulfide layer was worn away gradually, and the content of S element on the surface became reduced as well. After 2min, there was almost no S element on the sulfurized surface, and scuffing had occurred. Figure 2.52 shows the morphology of wear scar of steel ball rubbing against the sulfurized surface worn about 26s, and its distribution map of S element. It can be seen that almost no wear occurred on the surface of steel ball, and a large black sulfide film adhered to it.

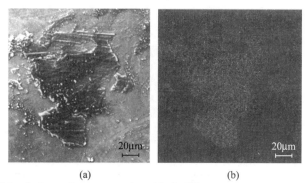

(a)   (b)

Fig.2.52   (a) Morphology of wear scar of steel ball sliding against 52100 steel sulfurized surface and (b) sulfur distribution map on it

The morphologies of wear debris from sulfurized surface and the plain surface are shown in Figure 2.53. It can be found that the wear debris of sulfurized surface at the time of 10s showed plate-like shape, and the roll-up or fold-up could be observed at their edge. These features were generated by the ploughing action of hard asperities against a soft surface during running-in period. At the time of 26s, the wear debris showed the feature of microchip-like delamination. The large plate-like wear debris were observed for the plain surface at the time of 18s, including plenty of spherical particles. The formation of spherical debris was regarded as the result of strain-fatigue wear or too high temperature induced by friction.

(c)

Fig.2.53   Morphologies of wear debris from 1045 steel sulfurized surface at the time of (a) 10s, (b) 26s and (c) from the plain surface at the time of 18s

2.4.1.4   Tribological properties of sulfuration layer under paraffin lubrication

The variation of friction coefficient μ with time under paraffin lubrication condition was almost the same as that under dry condition. The friction coefficient μ_b at initial stage, the endurance life t_c (when $\mu = 0.3$) of the layer, and the μ_t at final stage were considered as the indexes to evaluate the tribological behavior of sulfide layer. In the process of test, the friction coefficients of all the samples were in the range of 0.08—0.10, and were independent of the substrates and surface conditions. The μ_t at final stage is shown in Figure 2.54. The μ_t of sulfuration layer was still lower than that of unsulfurized surface. Figure 2.55 is the variation of endurance life t_c with the thickness of sulfide layer. The t_c of unsulfurized surface was very short, while that of sulfuration layer extended apparently and increased with the rise of thickness. Figure 2.56 shows the variation of wear scar width with layer thickness. The wear scar width of sulfurized layer was obviously less than that of unsulfurized surface, and almost unchanged with layer thickness. Moreover, in the same experimental conditions and surface states, the t_c of sulfurized layer on 1045 steel was shorter than that on 52100 steel and the w of sulfurized layer on the former was also more than that on the latter. Figure 2.57 shows the morphologies of wear scars as worn about 4.5min at 50N, 15Hz, 1mm under paraffin oil lubrication condition. It can be seen that the wear scar of the plain surface was rather rough, and scuffing had taken place on that of 52100 steel while the wear scar of the sulfurized surface was still fine and no scuffing occurred. The results of EDX analysis in Figure 2.58 show that some sulfur still remained on the wear scar of sulfurized surface.

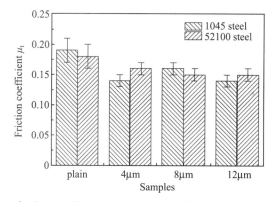

Fig.2.54   The μ_t at final stage of the samples under paraffin lubrication condition

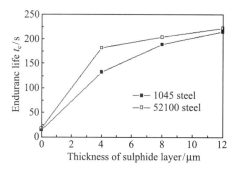

Fig.2.55   Variation of endurance life t_c with the thickness of sulfide layer under paraffin lubrication condition

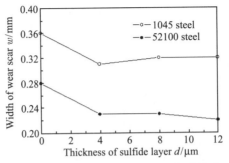

Fig.2.56 Variation of wear scar width with the thickness of sulfide layer under paraffin lubrication condition

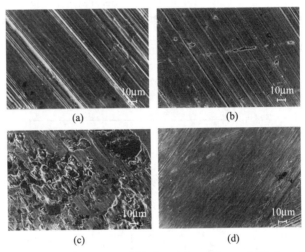

(a)   (b)

(c)   (d)

Fig.2.57 Morphologies of wear scars of (a) the plain surface and (b) the sulfurized surface with a 12μm thick layer of 1045 steel, (c) the plain surface and (d) the sulfurized surface with a 12μm thick layer of 52100 steel as worn about 4.5min at 80N, 15Hz, 1mm under paraffin oil lubrication condition

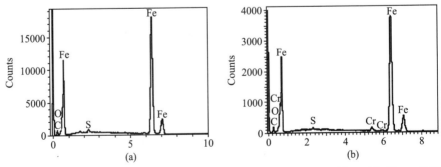

(a)   (b)

Fig.2.58 Composition of the wear scar of sulfurized surface on (a) 1045 steel in Figure 2.57 (b) and (b) 52100 steel in Figure 2.57 (d)

2.4.1.5   Tribological properties of sulfurized layer under engine-oil lubrication condition

1. Anti-scuffing performance of sulfide layer

Figure 2.59 shows the P-v diagram of the sulfurized and plain surface under engine-oil lubrication condition. The anti-scuffing loads for the sulfurized surface were considerably higher than those for the plain surface at low sliding speeds, especially, the specimen with 8μm thick sulfide layer showed the highest anti-scuffing load. However, the sulfide layer did not exhibit obvious anti-scuffing behavior at relatively high sliding speed.

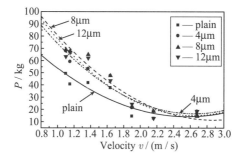

Fig.2.59 P-v diagram of the sulfurized and plain surface under oil lubrication condition

2. Friction-reducing performance of sulfide layer

Figure 2.60 shows the variations of friction coefficient μ with load of the sulfurized

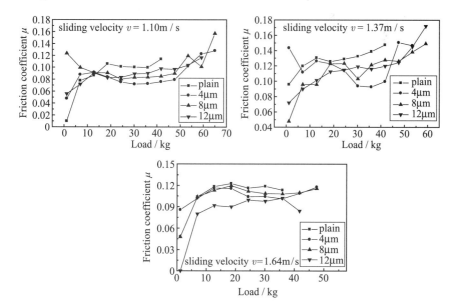

Fig.2.60   Variations of friction coefficient μ with load of the sulfurized and plain surfaces at sliding speed of 1.10m/s, 1.37m/s and 1.64m/s

and plain surfaces at sliding speed of 1.10m/s, 1.37m/s, 1.64m/s, respectively. The friction coefficients of sulfurized surface were all lower than that of plain surface.

3. Wear-resistance of sulfide layer

Figure 2.61 shows the width of wear scar on the disc and their counterpart ball after the wear experiments. It can be seen that the widths of wear scars on the sulfurized discs and balls, especially with 8μm layer, were smaller than those on the plain specimen.

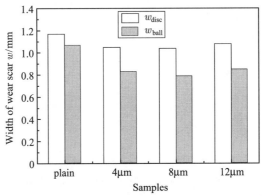

Fig.2.61   Widths of wear scars on the discs and their counterpart balls after the wear experiments

Figure 2.62 shows the cross-section profiles of wear scars measured by a surface profilometer, the data in the figures represent the depth of wear scar. Based on the wear scar width and depth, the wear volumes were calculated, shown in Table 2.5. Although the wear volume of 12μm thick sulfide layer was relatively larger, the wear scar depth was basically within the layer thickness. That means, the sulfide layer could protect the substrate and obviously improve the wear-resistance of the steel surface.

4. Composition of the boundary film

Figure 2.63 shows the morphologies of wear scars of the plain and sulfurized surfaces, and their counterpart balls. The wear scar of the plain surface was rougher than that of the sulfurized surface. The EDS analysis results are listed in Table 2.6. Besides oxygen, some remained sulfur was detected on the sulfurized disc, and on the ball surface as well, that means, sulfur was transferred to the counterface in the friction process. The content of oxygen on the ball surface was much more than that of the

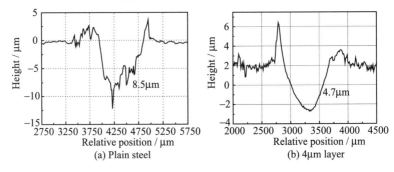

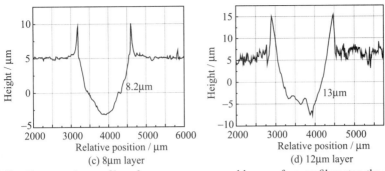

(c) 8μm layer

(d) 12μm layer

Fig.2.62   Cross-section profiles of wear scars measured by a surface profilometer, the data in the figures represent the depth of wear scar

Table 2.5   Wear scar widths and depths and wear volumes of the samples

Samples	Wear scar width w/mm	Wear scar depth d/μm	Wear volume V/mm³
Plain steel	1.17	8.5	0.896
4μm thick sulfide layer	1.05	4.7	0.458
8μm thick sulfide layer	1.04	8.2	0.750
12μm thick sulfide layer	1.08	13	1.169

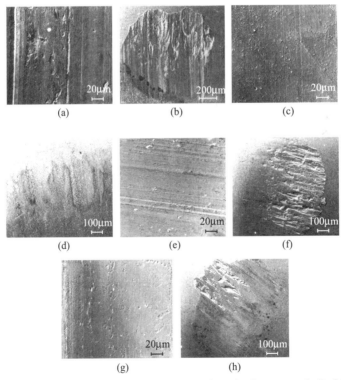

(a)   (b)   (c)

(d)   (e)   (f)

(g)   (h)

Fig.2.63   Morphologies of wear scars on (a) Plain surface, (b) Counterpart ball of (a), (c) 4μm sulfuration layer, (d) Counterpart ball of (c), (e) 8μm sulfuration layer, (f) Counterpart ball of (e), (g) 12μm sulfuration layer, (h) Counterpart ball of (g)

Table 2.6   Composition of the wear scar on each sample

Samples		Plain surface	4μm layer	8μm layer	12μm layer
	atom%S	—	0.67	0.57	0.11
Disc	atom%O	6.12	1.64	0.92	6.29
	atom%Fe	92.88	97.69	98.51	93.60
	atom%S	—	0.10	0.46	0.33
	atom%O	0.06	4.89	3.72	4.42
Ball	atom%Fe	98.84	93.02	95.19	94.35
	atom%Cr	1.10	2.00	0.63	0.90

ball against plain surface. The XPS analysis results are listed in Table 2.7. The content of boundary film of the plain surface was iron oxide, and that of the sulfurized surface was composed of iron oxide and FeS, the FeS_2 produced in the original sulfurized layer had disappeared. Figure 2.64 shows the composition profiles along depth of the discs. The content of oxide in the surface film was increased with the increase of thickness of sulfide layer, but the content of sulfide decreased.

Table 2.7   XPS analysis results of the wear scar on each disc sample

Discs	Bonding energy/eV			Compounds
	S (2p)	O (1s)	Fe (2p)	
Plain steel		531.4, 529.85	709.2	absorbed O_2, FeO or Fe_2O_3,
4μm layer	168.2, 162.4, 161.3	531.2, 529.5	710.8, 709.8	absorbed O_2, FeO or Fe_2O_3, S, $FeSO_4$, FeS
8μm layer	168.2, 162.3, 160.5	531.8, 529.4	711.0, 709.7	absorbed O_2, FeO or Fe_2O_3, S, $FeSO_4$, FeS
12μm layer	168.2, 162.3, 160.2	532.0, 529.7	711.0, 709.9	absorbed O_2, FeO or Fe_2O_3, S, $FeSO_4$, FeS

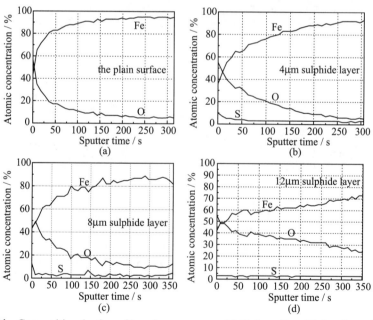

Fig.2.64   Composition depth profiles of wear scars on (a) Plain surface (b) 4μm layer (c) 8μm layer (d) 12μm layer analyzed by AES

(Sputtering rate was 30nm/min)

2.4.1.6   Contact fatigue behavior of sulfurized layer on 1045 steel

The sulfide layer was prepared on 1045 steel after quenching and tempering, and its fatigue life was tested on a contact fatigue tester of model JP-52. The experimental results showed that the fatigue life of the sulfurized layer was increased by a factor of 2, compared with that of unsulfurized surface.

Table 2.8 shows the data of contact fatigue life of sulfurized layer and plain surface, and their σ-N plot was shown in Figure 2.65.

Table 2.8   Data of contact fatigue life of sulfurized layer and plain surface

Contact pressure σ/MPa	Fatigue life N (unsulfurized)	Fatigue life N (sulfurized)	Increased range/%
4245	1.0558×10^6	1.8749×10^6	77.6
4859	0.7812×10^6	1.6439×10^6	110.4
5348	0.6988×10^6	1.3171×10^6	88.5
5760	0.6292×10^6	1.2029×10^6	91.2

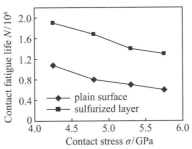

Fig.2.65 σ-N plot of sulfurized and plain 1045 steel

After sulfurizing, the contact fatigue life of the steel was improved obviously. The increased range was from 77.6% to 110.4%. With the contact pressure increasing, the contact fatigue life was shortened.

The basic process of contact fatigue is as follows: The surface and sub-surface produced accumulative shearing strain under the contact pressure, which caused the fatigue damage at the surface of steel. When the strain reached the critical value, the fatigue cracks were initiated, then the cracks propagated along the direction at a certain angle to the surface under the shearing stress, and finally, the rupture happened along the crack root under the normal stress to form spalling pits. When the samples were sulfurized, their surface became softened and easily produced plastic flow. Therefore, the actual contact area was expanded; that is, the actual contact pressure was descended, which was beneficial to the improvement of fatigue life. In addition, the surface roughness of unsulfurized sample was higher than that of sulfurized sample due to the effect of processing, the surface pressure regulation could greatly influence the distribution of stress field; consequently, the fatigue life of plain sample was shortened.

2.4.1.7   Friction-reducing and wear resistance mechanism of sulfurized layer

1. Wear process of sulfurized layer

The wear life of a solid lubrication film can be divided into two stages: before and after the occurrence of metallic contact. At the first stage, the friction occurs inside the solid lubrication film, thus the friction coefficient is very low, and the wear takes place basically inside the films. At the second stage, the rubbing surfaces come into direct contact, thus the severe wear will take place. At initial stage, sulfide layer was squashed and adhered to the surface of the counterpart. It impeded the metallic contact between rubbing surfaces. As the test going on, the sulfide layer was worn away, and friction coefficient went up due to the direct contact between metallic surfaces.

The friction-reducing mechanism of the sulfide layer can be shown by a scheme in Figure 2.66. During the process of ion sulfurization, the peaks of asperities on the surface had a greater probability of being bombarded by ions, so the sulfation depth was deeper than that at the valleys. The asperities on the surface were softened during the sulfuration process. This avoided the early scuffing and was favorable to the running-in process. During the running-in period, the valleys could be filled up with iron sulfide, and the real contact area was enlarged, which resulted in the decrease of contact stress. Therefore, the wear rate was reduced and the wear-resistance of the surface was improved. Furthermore, the sulfide layer could easily absorb lubrication oil and form an oil film because it was porous and loose. As the test was going on, the sulfide layer was worn away gradually, but there was still some sulfide remaining on the surface to play the role of solid lubrication.

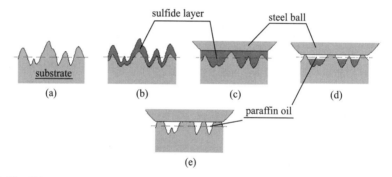

Fig.2.66   Schematic diagram of the friction-reducing effect of the sulfide layer

(a) The plain surface. (b) The sulfide layer produced on the surface by ion sulfuration. (c) The sulfide layer was squashed and adhered to the surface at the initial stage of wear test. (d) The sulfide layer was worn away, metallic surfaces came into direct contact, but some sulfur still remained on the surface. (e) The plain surface without sulfide layer came into direct contact and was severely worn

The sulfide layer could be decomposed and regenerated constantly due to the friction heat and contact pressure and it could exist on the frictional surface for a long time. Therefore, there was still some sulfide remaining on the surface after the test, which also made the final friction coefficient μ_t of the sulfide layer lower than that of plain surface. The endurance life t_c of sulfide layer was obviously influenced by the substrate, especially at high loads. Probably it was due to the less development of plastic deformation on the surface and the stronger support to the layer from the hard substrate. This point can be observed also in Figure 2.67, t_c (52100 steel)/t_c (1045 steel) was increased with increasing of load.

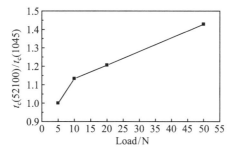

Fig.2.67   The ratio of endurance life of sulfide layer on 52100 steel t_c (52100) to that on 1045 steel t_c (1045) vs. load

2. Synergistic effects of sulfide and oxide on the frictional surface

As an excellent solid lubricant, iron sulfide possesses a high melting point (1100°C), very low hardness and shearing strength, and good adhesion to iron and steel surface. Therefore, sulfide layer can protect rubbing surface from seizure, and adhere to the counterface to reduce friction and hinder scuffing. Under dry friction condition, because the tests were conducted in air, oxide film could be formed. The existence of oxygen had a significant effect on the load-bearing property and wear-resistance. Organic sulfide almost can not exhibit anti-wear and extreme pressure property in the absence of oxygen or oxide film, it is more effective when there is Fe_3O_4. Research demonstrates that the friction coefficient depends on the average shear stress and load between the frictional surfaces, namely:

$$\mu = 2.276\bar{\tau} \left[\left(\frac{1-\nu_1^2}{\pi E_1} + \frac{1-\nu_2^2}{\pi E_2} \right) \frac{R}{\sqrt{P}} \right]^{2/3}$$

where μ is friction coefficient; $\bar{\tau}$ is average shear stress; ν, E are poisson and elastic modulus respectively; R is the radius of the ball and P is load. With the increasing of load, the friction heat increased and more oxides could be formed. The sulfides and oxides on the friction surfaces can further effectively prevent the direct contact between the metals, and the shear mainly occurred within the film. Consequently, the average shear stress between the frictional surfaces was almost invariant or decreased. The theoretical studies on the friction characteristics of sulfide layer on the iron and steel by some researchers also proved this point. Under the combined effect of sulfides and oxides, the final friction coefficient μ_t of sulfide layer was lower than μ of plain surface. The gap between μ_t and μ on 1045 steel was larger than that on 52100 steel (see Figure 2.38), which indicates that the sulfide layer on 1045 steel was easier to promote the formation of oxides on the frictional surface.

Under oil lubrication condition, the sulfide layer can be damaged not only by the mechanical action, but can also be decomposed and regenerated due to the contact pressure and frictional heat. The decomposed activated sulfur atoms react with iron to generate iron sulfide again, simultaneously, iron oxide is also produced. They can together play the role of boundary lubrication. The variation of concentration ratio of oxygen to sulfur along depth on the rubbing surface of sulfurized disc is shown in Figure 2.68. It can be found that the thicker the sulfide layer, the higher the concentration ratio of oxygen to sulfur. This indicates that sulfide layer can promote the formation of oxides, probably because of the loose structure and its high defect density, which is favorable for the diffusion of oxygen. In addition, the standard free energies of formation of FeO, Fe_2O_3, Fe_3O_4, FeS, FeS_2 are $-179.1kJ\cdot mol^{-1}$, $-742kJ\cdot mol^{-1}$, $-1016kJ\cdot mol^{-1}$, $-97.57kJ\cdot mol^{-1}$, $-166.69kJ\cdot mol^{-1}$, respectively. Sulfides can be readily displaced by oxides even with static exposure to oxygen because iron oxide is thermodynamically more stable than the iron sulfide. Furthermore, according to the theory of negative ions, the existence of oxides is indispensable for sulfide formation on friction surface. However, the formation of oxides can also hinder the growth of iron sulfide because of the competitive growth between them. Therefore, hardly any ferrous sulfide appears on the 12μm sulfurized surface initially, but much more ferrous sulfide appears on the rubbing surface of 4μm sulfide layer.

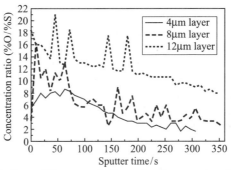

Fig.2.68 Depth distributions of the concentration ratio of oxygen to sulfur O/S on the rubbing surface of sulfurized discs analyzed by AES (Sputtering rate was 30nm/min)

Research shows that oxide film on the rubbing surfaces can not only catalyze the tribo-chemical reactions of sulfur-containing additives with the metallic surfaces, but also enhance the load-bearing capacity of sulfur extreme-pressure lubrication film and improve its anti-wear performance. A necessary condition for increasing the extreme pressure performance and reducing friction and wear is an optimum concentration ratio of O/S reaching 0.4—0.8 in the surface film.

2.4.2   Tribological Properties of Sulfide Layer on Four Kinds of Steel

2.4.2.1   Tribological properties of sulfurized layer under dry friction condition

Figure 2.69 shows the variation of friction coefficient with time of four sulfurized steels. Whether sulfurized or not, the variation laws of all steels were similar [3—5]. At beginning, the friction coefficient was low. After a short time of running, the friction coefficient increased quickly. The sulfurized layers were soon damaged under the severe dry condition due to their thin thickness and the coefficients reached a stable high value at 5min and increased gradually. While the plain steels could produce a boundary film on their surface through the reaction with additives in lubricating oil, which prevented the direct contact between the metals to a certain extent at the initial stage. However, the film was destroyed immediately and the friction coefficient increased as well. Comparing Figure 2.69 (a) and (b), it can be found that the friction coefficients of all sulfurized steels were obviously lower than those of unsulfurized steels.

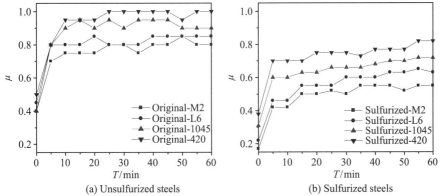

(a) Unsulfurized steels   (b) Sulfurized steels

Fig.2.69   Variation of friction coefficient with time under dry friction

Figures 2.70 and 2.71 are the variations of wear scar width with time and load respectively. The variation laws of the wear scar widths were also similar. With the increase of time and load, the wear scar widths increased, but the increase amplitudes of sulfurized steels were slower. The wear scar widths of sulfurized steels were always lower than those of original steels.

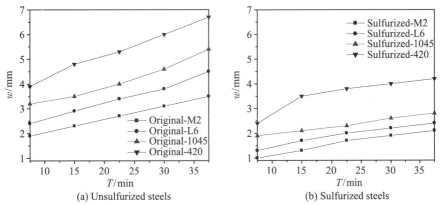

(a) Unsulfurized steels   (b) Sulfurized steels

Fig.2.70   Variation of wear scar width with time under dry friction

Figure 2.72 (a)—(d) show the worn morphologies of 4 sulfurized steels after sliding 30min under dry condition. All the sulfurized steels were damaged greatly, but the wear degree of sulfurized M2 steel was relatively the lightest, and some furrows existed on the worn surface, indicating it was mainly subjected to the abrasive wear; in addition to furrows, there were still spallings and corrosion pits on the worn surface of sulfurized L6 steel. Therefore, the main wear mechanisms of sulfurized L6 steel were abrasive and corrosive wear; there were not spallings and corrosion on the worn surface of sulfurized 420 steel, but severe plastic deformation occurred due to the low hardness of 420 steel. Therefore, the main wear mechanism of sulfurized 420 steel was abrasive wear.

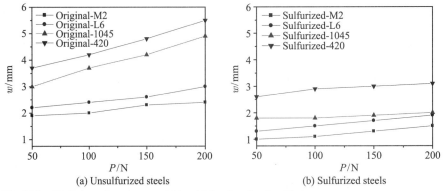

Fig.2.71   Variation of wear scar width with load under dry friction

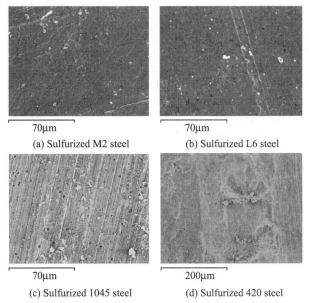

(a) Sulfurized M2 steel   (b) Sulfurized L6 steel

(c) Sulfurized 1045 steel   (d) Sulfurized 420 steel

Fig.2.72   Worn morphologies of the 4 sulfurized steels after sliding 30min under dry condition

2.4.2.2   Tribological properties of sulfurized layers under engine-oil lubrication condition

Figure 2.73 shows the variation of friction coefficient with time under oil lubrication condition. It can be found that the variation of friction coefficient of all steels was very smooth, indicating that under oil lubrication all friction-pairs were sliding at a quite steady state. The friction coefficients of all sulfurized steels were obviously less than those of unsulfurized steels due to the "combined-lubrication" effect.

Figures 2.74 and 2.75 show the variation of wear scar width with load and time respectively under oil lubrication condition. The wear scar widths of all sulfurized steels were decreased substantially, compared with those of the original steels, and they had the similar variation laws.

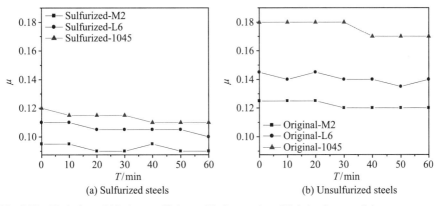

Fig.2.73   Variation of friction coefficient with time under oil lubrication condition

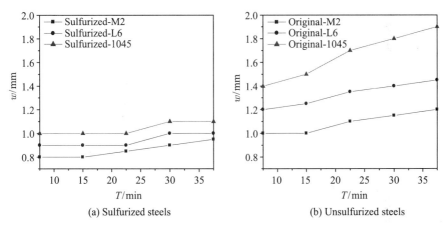

Fig.2.74   Variation of wear scar width with time under oil lubrication condition

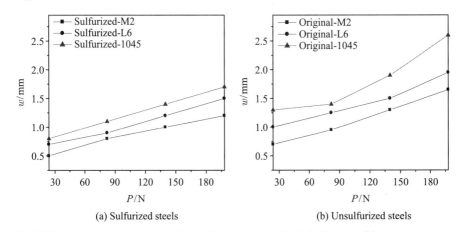

Fig.2.75   Variation of wear scar width with load under oil lubrication condition

Figure 2.76 shows the worn morphologies of the sulfurized steels after sliding

30min and the EDS analysis results of them. More complete pieces of film were still on the worn sulfurized M2 steel surface (Figure 2.76 (a)). Figure 2.76 (b) indicates that relatively higher sulfur content was still on the worn surface. It can be concluded that the sulphide layer on M2 steel possessed excellent friction-reducing behaviour and long service life. Figure 2.76 (c) shows obvious furrows and spalling pits on the worn sulfurized L6 steel surface, the wear-resistance of the sulfide layer on L6 steel was lower than that on M2 steel. The sulfide layer on 1045 steel had been damaged entirely, as shown in Figure 2.76 (e) and (f).

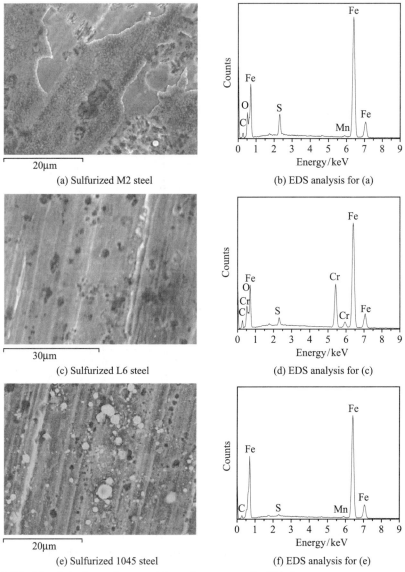

(a) Sulfurized M2 steel

(b) EDS analysis for (a)

(c) Sulfurized L6 steel

(d) EDS analysis for (c)

(e) Sulfurized 1045 steel

(f) EDS analysis for (e)

Fig.2.76 Morphologies and compositions of wear scar of sulfide layers

Figure 2.77 shows the variation of anti-scuffing load with velocity under oil lubrication condition. The anti-scuffing properties of all three steels, whether sulfurized or not, were relatively higher at low velocity. With increasing of velocity, the anti-scuffing property decreased. When the velocity exceeded 1.87m/s, the property deteriorated severely. More friction heats were produced at high velocity, which accelerated the decompositions of the sulfide layers. Consequently, the films lost their anti-scuffing properties.

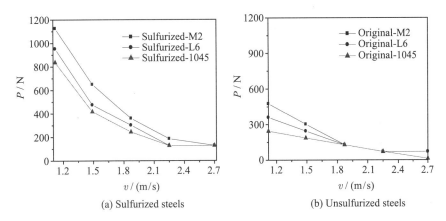

(a) Sulfurized steels   (b) Unsulfurized steels

Fig.2.77   Variation of anti-scuffing load with velocity under oil lubrication condition

Figure 2.78 shows the variation of friction coefficient with load at velocity of 1.12m/s and 1.49m/s. With the increase of load, the friction coefficient increased. The friction-reducing performance of sulfurized steels was always superior to that of the unsulfurized steels evidently.

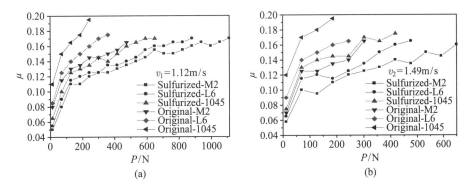

Fig.2.78   Variation of friction coefficient with load at a velocity of (a) 1.12m/s and (b) 1.49m/s

In a word, FeS solid lubrication film could be formed on various steels treated by low temperature ion sulfuration. FeS film could play the role of friction-reduction under dry condition, and exhibit superior friction-reducing property under oil lubrication through the "liquid + solid" combined lubrication mode.

2.4.2.3 The reasons for different tribological properties of sulfurized layers on different steels

The steels were treated by ion sulfuration at the same condition, but different steels showed different tribological properties [6]. The reasons might be related with the hardness, microstructure and corrosion resistance of substrates.

1. The influence of hardness

According to the tribological theory, an ideal friction surface should be soft at the surface, possessing excellent lubricating property, and hard at subsurface, that is, the hard substrate can give a sufficient support to the lubrication layer and keep a longer time of its effect. The hardness of sulfuration layer is very low, while the hardness of high speed steel is the highest among the three steels, reaching HRC 65, therefore its friction coefficient and wear volume are the lowest. The hardness of die steel and 1045 steel was relatively lower (HRC 55—58), the support to sulfide layer was weak. The film was broken rapidly and lost its friction-reduction and wear-resistance effects. The influence of hardness can be illustrated by the physical model shown in Figure 2.79.

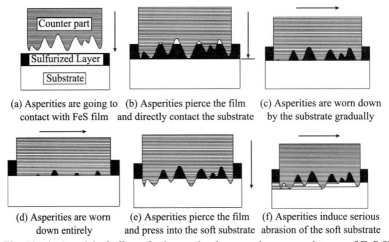

(a) Asperities are going to contact with FeS film (b) Asperities pierce the film and directly contact the substrate (c) Asperities are worn down by the substrate gradually

(d) Asperities are worn down entirely (e) Asperities pierce the film and press into the soft substrate (f) Asperities induce serious abrasion of the soft substrate

Fig.2.79　Physical model of effect of substrate hardness on the wear-resistance of FeS film

The friction between 52100 ball and sulfurized layer is actually the mutual friction between the multi-asperities on the ball surface and the sulfide layer as shown in Figure 2.79 (a). In the beginning, the steel ball asperities pierced the film and directly contacted with the substrate (see Figure 2.79 (b)). If the hardness of substrate was greater than that of the steel ball, asperities would be worn down by the substrate, as shown in Figure 2.79 (c) and (d). However, if the hardness of substrate was lower than that of the steel ball, asperities would continue to press into the substrate with a certain depth (see Figure 2.79 (e)). Asperities as a hard abrasive material induce serious abrasion (furrows and micro-cutting) of soft substrate, as shown in Figure 2.79 (f). Finally, the FeS film was destroyed under the normal stress and friction.

2. Influence of microstructure

High-speed steel (M2) contains a lot of carbide-forming elements, such as tungsten, molybdenum, chromium, and vanadium. They can produce the superhard carbides, like W_2C, $Cr_{23}C_6$, VC and so on with carbon. Die steel (L6) also contains carbide-forming elements of Cr, Mn, which can produce the hard carbides of $(Fe, Mn)_3C$, $(Fe,Cr)_3C$. These hard phases were likely to enhance the resistance of substrates to abrasion through dispersion hardening. Figure 2.80 (a) shows the cross-section morphologies of worn sulfurized die steel. The white grains are hard phases, whose micro-hardness is as high as $HV_{0.05}$ 1198. Figure 2.80 (b) shows the cross-section morphologies of worn sulfurized 1045 steel. 1045 steel does not contain alloying elements, and therefore the wear-resistance of sulfide layer on 1045 steel was lower than that on M2 and L6 steels.

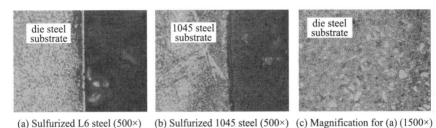

(a) Sulfurized L6 steel (500×)   (b) Sulfurized 1045 steel (500×)   (c) Magnification for (a) (1500×)

Fig.2.80   Morphologies of worn cross-section of samples

2.5   Influencing Factors of the Microstructures and Tribological Properties on Sulfurized Layers

2.5.1   Effect of the Substrate State on the Sulfide Layer on 1045 Steel

The geometry of 1045 steel testpiece was 24mm in diameter and 5mm in thickness. Its surface roughness was R_a =0.11mm. They were heat-treated by five different regimes: 860°C quenched (Q), 860°C quenched + 250°C tempered (QL), 860°C quenched + 400°C tempered (QM), 860°C quenched + 600°C tempered (QH), and 840°C annealed (A). Their microstructure, internal stress and hardness are listed in Table 2.9.

Table 2.9   Elementary situation of 1045 steel substrates of five different states and the sulfide layers on them

Testpiece No.		1#	2#	3#	4#	5#
Pretreatment		Q	QL	QM	QH	A
Microstructure		Martensite	Tempered martensite	Temper troostite	Temper sorbite	Ferrite and pearlite
Internal stress/(kg/mm²)		−114.95	−55.28	−44.78	−36.82	−21.23
Rockwell hardness/HRC before sulfurization		55—57	43—46	34—36	27—29	8—10
Rockwell hardness/HRC after sulfurization		50—52	44—46	34—36	28—30	7—10
Thickness of the sulfide layer/μm		6—7	5—6	3—5	2—4	2—3
Composition of the sulfide layer	atom%Fe	66.5	68.3	68.5	68.2	68.7
	atom%S	33.5	31.7	31.5	31.8	31.3

2.5.1.1   Effect of substrate state on the morphology and composition of sulfide layers

It can be seen from Table 2.9 that the sulfurization did not show an influence on the hardness of substrate at 200°C sulfurization temperature except a little decrease of the hardness of testpiece $1^{\#}$. The thickness of sulfide layer was increased in the sequence of $5^{\#} \rightarrow 1^{\#}$. Figure 2.81 shows the surface morphology of sulfide layer on testpiece $1^{\#}$. The other sulfide layers had obviously no difference with it. The microstructure of sulfide layer was loose and its grains showed the feature of clusters. The surface composition of sulfide layer listed in Table 2.9 was independent of the state of substrate. Figure 2.82 shows the cross-section morphologies of sulfide layers on testpiece $1^{\#}$ and $5^{\#}$. The layer thickness was homogeneous and no obvious transition zone could be found between the sulfide layer and substrate.

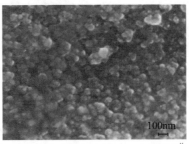

Fig.2.81   Surface morphology of the sulfide layer on testpiece $1^{\#}$

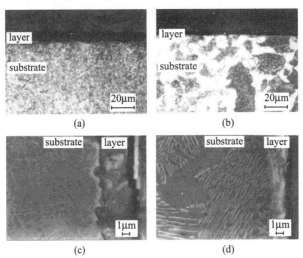

Fig.2.82   Cross-section morphologies of the sulfide layers on testpiece (a) $1^{\#}$ and (b) $5^{\#}$

2.5.1.2   Effect of substrate state on the phase structure of sulfide layers

Figure 2.83 shows the phase structures of sulfide layers on different substrates. Three phases of cubic α-Fe, hexagonal FeS and orthorhombic FeS_2 presented in every sulphide layer. The integrate intensity I of FeS (110) peak and α-Fe (211) peak, as well as the rate I_{FeS}/I_{Fe} of them are listed in Table 2.10. The full widths at half maximum (FWHMs) of every α-Fe peak are shown in the same table. It shows that I_{FeS}/I_{Fe} was independent of the states of substrate. This means that the contents of FeS in the sulphide layers on different substrates were probably consistent. The FWHM of every α-Fe peak became broadened in the sequence of $5^{\#} \rightarrow 1^{\#}$ due to the following possible reasons of instrumental effects, crystallite size and lattice strain. In these cases, the instrumental effects were the same, and the broadening due to crystallite size $B_{crystallite}$ was much larger than that due to lattice strain B_{strain}, i.e., $B_{crystallite} \gg B_{strain}$ at small diffraction angles of 2θ less than $100°$. Therefore, the broadening was mainly due to crystallite size. The smaller the average crystallite size and the higher the density of defect, the broader the FWHMs of the diffraction peaks.

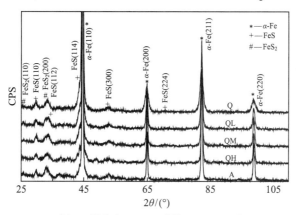

Fig.2.83   Phase structure of the sulfide layers on different state substrates

Table 2.10   Integrate intensities and FWHMs of FeS and α-Fe diffraction peaks

Testpiece No.	I_{FeS} (arb. unit) FeS(110)	I_{Fe} (arb. unit) α-Fe(211)	I_{FeS}/I_{Fe}	FWHM of α-Fe peak/($°$) (110)	(200)	(211)	(220)
$1^{\#}$	308.82	1521.87	0.203	0.56	1.34	0.96	1.19
$2^{\#}$	300.21	1603.52	0.200	0.33	0.56	0.56	0.68
$3^{\#}$	325.48	1603.25	0.203	0.28	0.40	0.49	0.67
$4^{\#}$	290.76	1633.60	0.178	0.28	0.40	0.48	0.48
$5^{\#}$	304.87	1714.52	0.178	0.26	0.37	0.45	0.61

2.5.1.3   Effect of substrate state on the tribological properties of sulfide layers

The variation of friction coefficient of sulfurized surface with time is shown in Figure 2.35. The friction coefficient μ_b at the initial stage, the endurance life t_c (when $\mu = 0.3$) of sulfide layer, and the μ_t at the final stage were regarded as the indexes to evaluate the friction-reducing property of sulfide layer. The width of wear scar after running of 30s and 2min was regarded as the index to evaluate the wear-resistance of sulfurized surface. The values of t_c, μ_b and μ_t are listed in Table 2.11. It can be seen that the μ_b of every sulfurized testpiece was almost similar in the range from 0.14 to 0.18, and was no more than one-third that of the plain surface. However, μ_t was increased in the sequence of $1^{\#} \rightarrow 5^{\#}$, as shown in Figure 2.84. Moreover, the μ_t of sulfurized surface was a little lower than that of plain surface. The t_c became shorter in the sequence of $1^{\#} \to 5^{\#}$. The width of wear scars w_1 at 30s and w_2 at 2min of different testpieces are shown in Figure 2.85. It can be seen that both w_1 and w_2 were increased in the sequence of $1^{\#} \to 5^{\#}$. Just like the behavior of t_c, it was related to the reduction of substrate hardness and the decrease of layer thickness. In addition, the w of all sulfurized surfaces was less than that of the plain surface. It means that the sulfide layer could improve the wear-resistance at the initial stage. After running

Table 2.11　The friction coefficient μ_b at the initial stage and μ_t at the final stage of sulfurized and plain surfaces, as well as the endurance life t_c of sulfide layer

Testpiece No.	μ_b	t_c/s	μ_t
$1^{\#}$—sulfurized	0.14—0.16	27	0.41—0.44
$1^{\#}$—plain	—	—	0.42—0.46
$2^{\#}$—sulfurized	0.16—0.18	23	0.41—0.51
$2^{\#}$—plain	—	—	0.54—0.57
$3^{\#}$—sulfurized	0.16—0.17	24	0.48—0.53
$3^{\#}$—plain	—	—	0.54—0.60
$4^{\#}$—sulfurized	0.16—0.17	20	0.51—0.56
$4^{\#}$—plain	—	—	0.59—0.63
$5^{\#}$—sulfurized	0.14—0.16	17	0.53—0.64
$5^{\#}$—plain	—	—	0.61—0.66

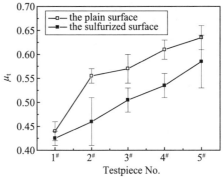

Fig.2.84　Final friction coefficient μ_t of the sulfurized and plain surfaces of different state substrates

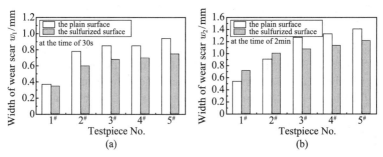

Fig.2.85　Width of wear scar (a) w_1 at the running time of 30s and (b) w_2 at 2min of the sulfurized and plain surface

of 2min, the w of sulfurized testpieces of $1^{\#}$ and $2^{\#}$ was larger than that of plain surface; however, the w of other sulfurized surfaces was still less than that of plain surface.

2.5.2 Effect of Environment Temperature on the Sulfurized Layer on 52100 Steel

2.5.2.1 Changes of structures and tribological properties of vacuum heated sulfide layer

The experimental material was 52100 steel after quenching and low temperature tempering, with a hardness of HRC 57—62. The geometry of testpiece was 24mm in diameter and 5mm in thickness. Its surface roughness was R_a =0.11mm. To study the change of sulfide layer at a high temperature, the sulfurized sample was heated in a vacuum furnace with a vacuum degree of 1.3×10^{-4}Pa, heating temperature of 180°C and 360°C, heating time of 0.5h and 1.5h.

Table 2.12 shows the hardness changes of the 52100 steel heated in vacuum. The substrate hardness decreased after a long time heating at 360°C. Figure 2.86 shows the cross-section morphology and elements distributions along cross-section of the sulfide layer heated for 1.5h at 180°C. The compositions of heated sulfide layers are shown in Table 2.13. The compositions of sulfide layers heated at different temperatures did not change significantly. Figure 2.87 shows the phase structures of sulfide layers heated for different times at different temperatures. The phase structures of heated sulfide layers did not show any change, and the layer was still composed of FeS, FeS_2, α-Fe, and Fe_5C_2. The above results indicated that the sulfide layers were stable at high temperatures in vacuum, whose compositions and structures did not change.

Table 2.12 Hardness changes of the 52100 steel heated in vacuum

Heating time	Hardness/HRC		
	Room temperature	180°C	360°C
0.5h	55.0	56.0	56.0
1.5h	55.0	56.5	47.0

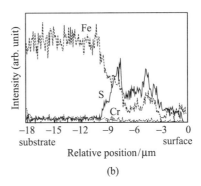

(a)   (b)

Fig.2.86 (a) Cross-section morphology and (b) elements distributions along cross-section of the sulfide layer heated for 1.5h at 180°C

Table 2.13 Compositions of sulfide layers heated for a certain time at different temperatures

Heating temperature and time	Atom%Fe	Atom%S
Original sulfide layer	54.32	45.68
180°C, 0.5h	55.66	44.34
180°C, 1.5h	52.73	47.27
360°C, 0.5h	52.78	47.22
360°C, 1.5h	53.14	46.86

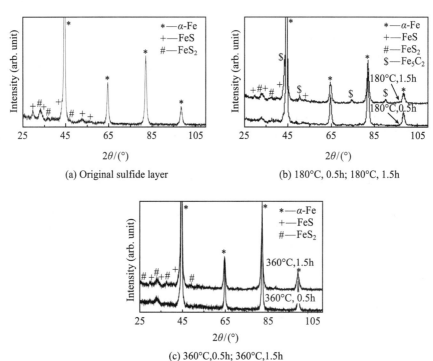

(a) Original sulfide layer

(b) 180°C, 0.5h; 180°C, 1.5h

(c) 360°C,0.5h; 360°C,1.5h

Fig.2.87 Phase structures of sulfide layers heated for different times at different temperatures

The variation of friction coefficient with time is shown in Figure 2.35. The friction coefficient μ_b at initial stage, the endurance life t_c (when $\mu = 0.3$) of the layer, and the μ_t at final stage were considered as the indexes to evaluate the tribological behaviors of sulfide layer. The values of μ_b, t_c and μ_t are shown in Table 2.14. The endurance life t_c of sulfide layer was shortened obviously at high temperatures, even $t_c = 0$ at 360°C. The friction coefficient μ_b of sulfide layer was slightly higher at 180°C than that at room temperature. The μ_t of sulfide layer slightly decreased, compared with that of unsulfurized steel, while there was no significant difference between them at high temperatures. In addition, the μ_t of all samples decreased gradually with increasing temperature. Figure 2.88 shows the wear scar widths of all samples. Obviously, the wear scar width of sulfurized surface was less than that of unsulfurized surface. With the temperature increasing, the wear scar width got larger. The $\varepsilon = w_{sulfurized}/w_{plain}$ was regarded as the indexes to evaluate the wear-resistance of sulfide layer. The bigger the ε, the better the wear-resistance. The values of ε are shown in Table 2.15. It was found that the wear-resistance of sulfide layer would be lower under high temperature.

Table 2.14 Tribological data of sulfurized and unsulfurized 52100 steel at different temperatures

		μ_b	t_c	μ_t
Room temperature	Sulfurized	0.18—0.20	20	0.43—0.50
	Unsulfurized	—	—	0.48—0.54
180°C	Sulfurized	0.20—0.22	3—5	0.42—0.50
	Unsulfurized	—	—	0.44—0.47
360°C	Sulfurized	—	0	0.35—0.37
	Unsulfurized	—	—	0.35—0.39

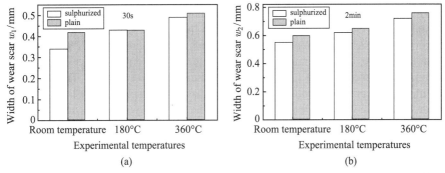

Fig.2.88 Wear scar width of sulfurized and unsulfurized steels after sliding (a) 30s and (b) 2min

Table 2.15 Relative wear-resistance ε of sulfide layers under different temperatures

	Room temperature	180°C	360°C
ε (30s)	1.24	1.00	1.04
ε (2min)	1.09	1.05	1.06

2.5.2.2 Changes of structures and compositions of sulfide layers heated in air

Table 2.16 shows the EDS analysis results of sulfide layer without wear. The content of oxygen increased significantly. Figure 2.89 shows the phase structures of sulfide layers heated for 30min in air at different temperatures. With the increase of temperature, the content of Fe_2O_3 became larger gradually.

The above experimental results indicated the sulfide layer heated in vacuum did not continue to diffuse along the substrate defects into inner and the phase structures

Table 2.16 Composition of sulfide layer heated for 30min in air at different temperatures

	Atom%Fe	Atom%S	Atom%O	Atom%Cr
Room temperature	74.08	21.66	2.99	1.28
180°C	73.37	20.99	4.24	1.40
360°C	70.05	20.56	8.65	0.72

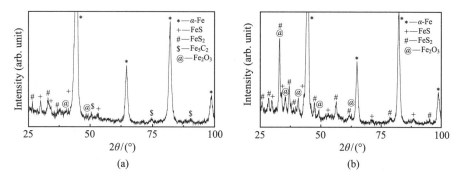

Fig.2.89   Phase structures of sulfide layers heated for 30min in air at (a) 180°C and (b) 360°C

and compositions did not show any changes, indicating that sulfide layers were stable at high temperatures. When the sulfide heated in air, iron sulfide would be oxidized to generate ferric oxide. In addition, due to the loose and porous of sulfide layer, the irons in the layer and substrate also could be oxidized and ferric oxides were produced within the layer and at the junction between the layer and substrate. This made the sulfide layer become discontinuous and prevented the interaction of the layer and substrate during the wear process. Consequently, the sulfide layer was not easy to adhere to the surfaces, resulting in the weak wear-resistance of the layer. Moreover, to a certain extent, ferrous oxide acting as an abrasive material could accelerate the wear of the layer. When the environment temperature increased, the content of generated ferrous oxide increased and the lubrication effect decreased; meanwhile, the endurance life and anti-friction behavior both descended. The practices showed that the life of the sulfurized roller with 12mm and 10mm diameter wire rods could be improved by a factor of 1—2, while the improvement was not significant for 8mm diameter. This might be related to the severe oxidation of sulfide layer.

2.5.3   Effect of Wear Conditions on the Tribological Behaviors of Sulfurized Layer on 52100 Steel

2.5.3.1   Effect of velocity

Figure 2.90 shows the friction coefficient and weight loss of sulfurized discs at different velocities. With the increasing velocity, the friction coefficient and weight loss of sulfide layer both increased.

2.5.3.2   Effect of load

Figure 2.91 shows the friction coefficient and weight loss of sulfurized discs at different loads. Under a certain velocity, with the load getting larger, there was significant linear decreasing of the friction coefficient. It decreased to the minimum at 300N and began to increase. The weight loss increased with increasing load, indicating that the wear of FeS was accelerated.

Although the increasing load could reduce the lubrication effect of FeS, under each velocity, there was always a load at which the FeS layer could fully take the role of friction-reduction. For instance, under the velocity of 800r/min, the friction coefficient was the lowest at a load of 300N.

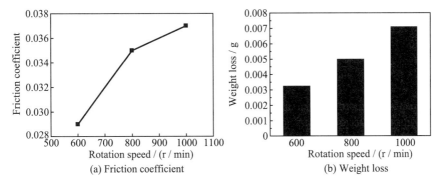

Fig.2.90 Friction coefficient and weight loss of sulfurized discs at different velocities

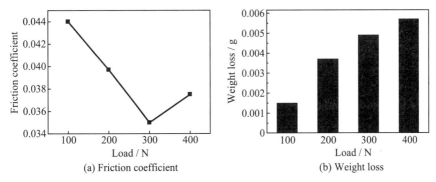

Fig.2.91 Friction coefficient and weight loss of sulfurized discs at different loads

References

1. Zhu L N, Li G L, Wang H D, et al. Comparative characterization to the three kinds of solid-shape FeS [J]. Materials Letters, 2008, 62(1–2): 163–166.
2. Wang H D, Zhuang D M, Wang K L, et al. Microstructure and wear behaviors of ion sulfide layer on high-speed steel [J]. Journal of Materials Science Letters, 2002, 21(19): 1545–1548.
3. Wang H D, Zhuang D M, Wang K L, et al. Anti-scuffing properties of ion sulfide layers on three hard steels [J]. Wear, 2002, 253(11–12): 1207–1213.
4. Wang H D, Xu B S, Liu J J, et al. Investigation on friction and wear behaviors of FeS films on L6 steel surface [J]. Applied Surface Science, 2005, 252(4): 1084–1091.
5. Wang H D, Xu B S, Liu J J, et al. Weak corrosion behavior on sulphurized steel surface during friction [J]. Applied Surface Science, 2005, 252(5): 1704–1709.
6. Wang H D, Zhuang D M, Wang K L, et al. The comparison on tribological properties of ion sulfuration steels under oil lubrication [J]. Materials Letters, 2003, 57(15): 2225–2232.

Chapter 3

FeS Solid Lubrication Film Prepared by a Two-step Method

The sulfide layer prepared by low temperature ion sulfuration technique has been introduced in Chapter 2. This chapter will give an account of new methods for preparing FeS film—two-step method, namely the precursor was firstly fabricated by sputtering, shot-peening, nitriding, carbonitriding, or spraying, and then was treated by ion sulfuration to form FeS film.

3.1 Radio-frequency (RF) Sputtering + Sulfurizing Combined Treatment

The nature of ion sulfuration technology is that the active sulfur atoms penetrate into the iron substrate and react with iron to form FeS. However, the technique is only applicable to the ferrous metal substrates, which can provide Fe atoms. In order to expand the application of ion sulfuration technique, the two-step method was created, i.e., the Fe film was first prepared on non-ferrous metal surface by RF sputtering, and then was treated by ion sulfuration.

3.1.1 RF Sputtering Technology

3.1.1.1 Principle of RF sputtering

(1) Glow discharge.

Glow discharge is a discharging phenomenon produced by applying DC voltage to the two electrodes in a vacuum chamber with pressure of 10^{-2}—10Pa. Figure 3.1 shows the volt-ampere characteristic curve of glow discharge: When the voltage in AB-step gradually increases from zero, a weak current (10^{-18}—10^{-16}A) is produced due to the electron emission and space ionizing caused by the cosmic radiation. The BC-step is a self-sustained dark discharge process, and the current is almost constant, it is the Townsend discharge with weak luminescence. The CD-step is a transition zone; the DE-step is normal glow discharge, the current has nothing to do with the voltage with bright glow between the two electrodes. The EF-step is abnormal glow discharge with the increasing of discharge voltage and current density. The FG-step is arc discharge with low decrease of the voltage and rapid decrease of the current. The current density of normal glow discharge is related to the cathode material, gas type, gas pressure, and cathode shape, etc. The re-sputtering and other glow discharge all work in the abnormal glow discharge zone.

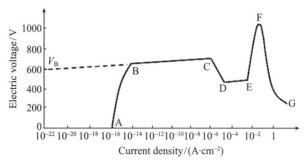

Fig.3.1   Volt-ampere characteristic curve of glow discharge

(2) RF sputtering.

If the alternating current is applied to the two electrodes and the frequency of radio-frequency is above 50kHz, the produced glow discharge is known as RF glow discharge; meanwhile, the sputtering using RF glow discharge is called RF sputtering. During sputtering, when the target is in the negative half-period, Ar^+ bombards the target with high kinetic energy in the electric field so that the atoms or molecules on the target surface are sputtered to form film on the substrate surface. When the target is in the positive half-period, at first, the electrons move to the target electrode and neutralize the positive charges on the target surface. As the migration rate of the electron is much more than that of the ion, in addition to the neutralization of the positive charges, a large number of electrons are gathered rapidly so that the target surface presents negative potential, and it can still attract Ar^+ to bombard the target to form film. The process of RF sputtering is shown in Figure 3.2.

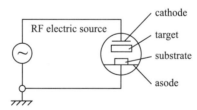

Fig.3.2   Scheme of the RF sputtering process

3.1.1.2   Characteristics of RF sputtering process

Compared with other sputtering methods, RF sputtering possesses the following characteristics.

(1) The applicable target materials have a wide range; the sputtering power can be adjusted conveniently.

As the RF voltage can couple through any kind of resistance, electrode is no longer limited to be the conductor. For the material with poor conductivity, when using RF sputtering to improve the deposition yield, we can avoid the occurrence of the substantial increase in voltage, while the phenomenon is ubiquitous by using other methods. Meanwhile, the RF power can be effectively inputted by adjusting matches of the discharge impedance and power impedance.

(2) The self-bias effect of target.

For insulated target material, the migration rate of the electron is much bigger than that of ion in the RF electric field. When colliding with the target, electrons will be accumulated on the target surface, and the target is automatically in a negative potential; the positive ions in the plasma region will be attracted by the negative ions on the target surface, so that the gas ions spontaneously bombard and sputter the target. If the target is a good electrical conductor, the power supply can be coupled to the target by connecting the capacitance in series so that the current path is isolated to form negative bias.

(3) The deposition rate is high.

As the effect of self-bias and oscillations in the high-frequency electric field increase the ionization probability of electrons, the sputtering rate can be improved.

3.1.2   Process of Preparation

The substrate material used was single crystal silicon (100) and its surface roughness RMS was 0.2nm. The substrate material for friction and wear test was 1045 steel, heat-treated by quenching and low temperature tempering. The hardness was HRC 52 and the surface roughness R_a was 0.13μm. The Fe film with thickness of 5μm was first prepared on the surfaces of single crystal silicon and 1045 steel by RF sputtering and then treated by low temperature ion sulfuration to form FeS film.

The Fe film was sputtered in the RF equipment of model MIP-800. The working gas was Ar (99.999% pure). The degree of background vacuum was set for 6×10^{-3}Pa, and the working pressure was 6×10^{-1}Pa. The sintered pure Fe target (99.99%) was used. In order to improve the deposition efficiency, the negative bias of 150V was applied on the surface. Before deposition, the substrate was heated to 100°C. The plate pressure and current were set up for 800V and 300mA respectively. Self-bias voltage was 750V and sputtering power was 200—220W. The film thickness was precisely controlled by the deposition of time. The deposition rate measured by AES was 11nm/min. The metal films with 5μm thickness were finally prepared.

Low temperature ion sulfuration was conducted in the sulfurizing equipment of model DW-1. The workpieces (single crystal silicon and 1045 steel coated with the Fe film) were put on the cathode tray and the container wall was linked to the anode. When the vacuum was up to a certain value, the ammonia gas was filled into the chamber. The high-tension direct current was applied between the cathode and anode to ionize the ammonia gas. Under the effect of cathode voltage drop, the ammonia ions were accelerated to bomb the cathode (workpieces), elevating their temperature. Till the temperature was up to a scheduled value of 190°C, the bombardment stopped. In this temperature the solid sulfur in the chamber became gasified, and the sulfur gas atoms permeated into the Fe film through the crystal defects and grain boundaries to produce the FeS film. The sulfurizing time was 2h.

3.1.3   Microstructures

The scanning electronic microscope (SEM) and atomic force microscope (AFM) were utilized to observe and analyze the morphologies of the surface, cross-section and worn scar, as well as the composition of the sulfide layer. X-ray diffraction (XRD) was used to analyze the phase structure of the Fe film and sulfide layer. The scratch tester of model WS-97 was used to measure the bonding strength between the FeS film and substrate. The nano-hardness and elastic modulus of FeS film were determined by the nano-indentation tester.

3.1.3.1   Phase structure

Figure 3.3 (a) shows the X-ray diffraction pattern of the Fe film. It has a polycrystalline structure and preferred orientation, indicating that the close-packed plane (110) was basically parallel to the sample surface. Figure 3.3 (b) is the XRD pattern of FeS film. FeS and FeS_2 were generated after sulfurizing, but Fe is still the main phase. It can be considered that the composite FeS film prepared by the two-step method was a metal-base composite film containing the solid lubricant FeS.

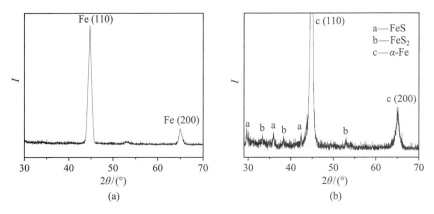

Fig.3.3   XRD patterns of Fe film (a) and FeS film (b)

3.1.3.2   Surface morphologies

Figure 3.4 shows the surface morphologies of Fe film and FeS film by AFM. The Fe

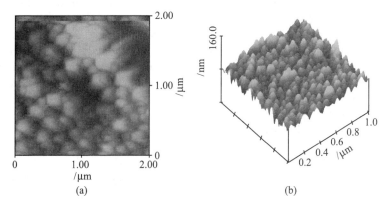

Fig.3.4   Surface morphologies of Fe film and FeS film by AFM (a) Fe film and (b) FeS film

film grew in the direction of the columnar crystal, while FeS film gradually evolved into sharp islets after sulfuration. This was due to that in the initial stage of sulfurizing, the ammonia ions had a certain impinging effect on the Fe film when the ammonia ions bombarded the Fe film.

3.1.3.3   Nano-mechanical properties

Table 3.1 shows the nanohardness and elastic modulus of iron piece, Fe film and FeS film. The nanohardness of Fe film was larger than that of iron piece. There was relatively high compressive stress on the surface of Fe film during sputtering so that its hardness increased. Meanwhile, the nanohardness and elastic modulus of the FeS film were much lower than those of Fe film. As the elastic modulus reflects the difficulty degree of the atoms leaving a balance position and relies on the atomic bonding force in the crystal; meanwhile, parts of Fe atoms have been seized by the S atoms to form FeS, the continuity of Fe atom was destroyed and the atomic bonding force decreased. Therefore, the elastic modulus of the FeS film decreased significantly.

Table 3.1   Nanohardness and elastic modulus of iron piece, Fe film and FeS film

Sample	Iron piece	Fe film	FeS film
Nanohardness/GPa	2.35	2.65	1.35
Elastic modulus/GPa	200.20	215.70	150.85

3.1.3.4   Bonding strength between the film and substrate

The measured results showed that the bonding strength between Fe film and substrate was up to 61.6N, indicating that the metal film bonds well with the substrate. However, the bonding strength between FeS film and substrate was only 51.7N, obviously lower than that of Fe film. This is due to the reaction of S and Fe, which destroyed the continuity of Fe atoms at the interface of the film and substrate.

3.1.4   Tribological Properties

The friction and wear tests were carried out on a ball-on-disc testing machine of model DD-92 under dry condition. As comparison, the plain 1045 steel was also tested under the same condition. The upper sample was the 52100 steel ball with a diameter of 8mm, hardness HV 770, and the lower sample was the 1045 sulfurized steel disc with dimension of ϕ45mm$\times$8.5mm. In operation, the upper sample was fixed and the lower sample rotated. The wear loss was expressed by the width of wear scar. Scuffing load was expressed by the weight of load when the film was initially scuffed. When testing the variation of friction coefficient with time, a load of 0.2N and velocity of 0.158m/s were fixed and the time changed from 0 to 60min. When testing the variation of worn scar widths with load, a time of 37.5min and velocity of 0.158m/s were fixed and the loads were 0.5N, 1.0N, 1.5N and 2.0N. When testing the variation of scuffing load with velocity, the following velocity variables were employed 0.104m/s, 0.158m/s, 0.209m/s and 0.262m/s. At each velocity, the sample ran in for 1min at first under loads of 0.1N and 0.2N, respectively. Then the load was added stepwisely. The increment of load was 0.5N and the duration time was 2min. When the frictional force suddenly increased and accompanied by evident oscillation and noise, this indicated the occurrence of scuffing, and the sum of all weights was the scuffing load.

Figure 3.5 shows the tribological properties of FeS film and 1045 steel. The friction coefficient of FeS film was only 1/3 times that of 1045 steel in the initial stage, while the coefficient of the former was nearly 1/4 times that of the latter in the final stage (Figure 3.5 (a)). This indicated that the lubrication role of FeS was getting more and more obvious with the friction time increasing, this was probably due to that FeS film transferred between the frictional pairs improved the lubrication condition. The wear scar width of FeS increased with increasing of load, but was obviously less than that of plain 1045 steel (Figure 3.5 (b)). The anti-scuffing load of FeS film was about 2.5 times larger than that of 1045 steel (Figure 3.5 (c)). With the increase of velocity, the anti-scuffing properties descended, indicating that the anti-scuffing property was very sensitive to the velocity. The generated friction heat was more at a high velocity, which promoted the decomposition of FeS film and the adhesion between the 1045 steel disc and 52100 steel ball.

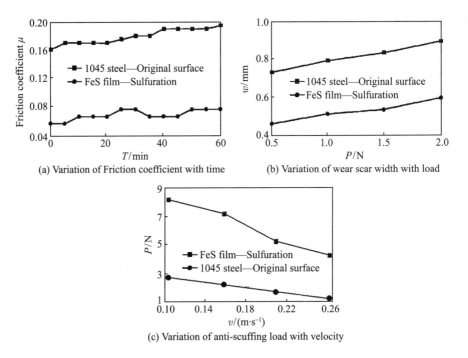

(a) Variation of Friction coefficient with time

(b) Variation of wear scar width with load

(c) Variation of anti-scuffing load with velocity

Fig.3.5 Tribological properties of the FeS film

Figure 3.6 shows the worn morphologies after scuffing of the FeS film and 1045 steel. The failure type of the FeS film was spalling, while that of 1045 steel was adhesion, obvious adhesion spots could be seen in Figure 3.6 (b). As the surface of 1045 steel lacked lubrication, the direct contact between metals and adhesion occurred under the frictional heat.

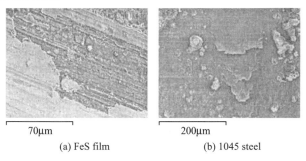

70μm	200μm
(a) FeS film	(b) 1045 steel

Fig.3.6 Worn morphologies after scuffing of the FeS film and 1045 steel

3.2 Shot-peening + Ion Sulfuration Combined Treatment

Shot-peening treatment can significantly improve the fatigue life and stress corrosion resistance of the workpiece, and it can also increase the rolling contact fatigue life under the low contact stress condition. Shot-peening can cause some changes of the metal surface as follows: ① The plastic deformation layer can be formed, and the dislocation density increases to $10^{12} \mathrm{cm}^{-2}$. Meanwhile, the sub-boundary and grain refinement phenomena take place. ② The residual compressive stress is generated. At the same time, the residual austenite is induced to a martensitic transformation, which can further increase the compressive residual stress of the surface. ③ The surface morphologies and roughness have changed. The research of the previous chapter has proved that the sulfur mainly diffused into the substrate along the defects during the sulfurizing process. Therefore, the high defect density on the shot-peened surface may accelerate the diffusion of the sulfur to thicken sulfide layer. Under the mixed and boundary lubrication conditions, the pits formed by shot-peening can increase the oil supply in the contact area so that the oil film is easily to be formed; meanwhile, the rough surface is beneficial to the attachment of solid lubrication film to postpone the occurrence of scuffing. Therefore, the composite treatment of shot-peening and sulfuration is helpful to make full use of sulfide layer and further improve the tribological properties of the material surface.

The 1045 and 52100 steels were shot-peened with different intensities, and then were treated by low temperature ion sulfuration. The structure and tribological properties of sulfide layer under dry and oil lubrication conditions were studied systematically. The wear mechanisms of sulfide layers were also discussed.

3.2.1 Preparation

Experimental materials were 1045 steel treated by quenching and high temperature tempering with a hardness of HB 174—197, and 52100 steel treated by quenching and low temperature tempering with a hardness of HRC 46—51. The average surface roughness R_a=0.11μm. The surfaces of samples were first treated by shot-peening with two intensities of 0.15mmA and 0.25mmA, and then were treated by ion sulfuration.

3.2.2 Characterization

3.2.2.1 Surface roughness, residual stress and hardness of the sulfide layer

The surface roughness R_a of shot-peening + sulfurizing and shot-peening on 1045 steel with two intensities are shown in Table 3.2. With the increasing of shot-peening intensity, the R_a increased. Table 3.3 shows the residual stress σ_r. It can be found that sulfurizing had little effect on the surface roughness and residual stress after shot-peening. The micro-hardness $HV_{0.1}$ after shot-peening + sulfurizing combined treatment is shown in Figure 3.7. As the shot-peening intensity became larger, both the surface hardness and hardening layer depth increased.

Table 3.2   Surface roughness R_a of 1045 steel after shot-peening + sulfurizing and shot-peening

$R_a/\mu m$	After shot-peening	After shot-peening + ion sulfurizing
0.15mmA	2.43	2.57
0.25mmA	3.66	3.00

Table 3.3   Residual stress σ_r of 1045 steel after shot-peening and shot-peening + sulfurizing

$\sigma_r/(kg/mm^2)$	After shot-peening	After shot-peening + ion sulfurizing
0.15mmA	-38.88	-37.70
0.25mmA	-27.39	-30.95

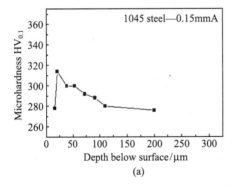 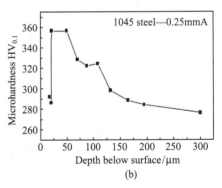

Fig.3.7   Distribution of the micro-hardness $HV_{0.1}$ along depth of 1045 steel treated by shot-peening with intensities of (a) 0.15mmA and (b) 0.25mmA + ion sulfurizing

3.2.2.2 Thickness of the sulfide layer

The cross-section morphologies of 1045 steel treated by sulfurizing and shot-peening with intensities of 0.15mmA and 0.25mmA + sulfurizing under light microscope are shown in Figure 3.8. The thicknesses of sulfide layers on 1045 steel and 52100 steel are listed in Table 3.4. The thickness of sulfide layer obtained by shot-peening + sulfurizing was larger than that by sulfurizing, and the sulfide layer got thicker with the increase of shot-peening intensity, indicating that the shot-peening surface promoted the process of sulfurizing.

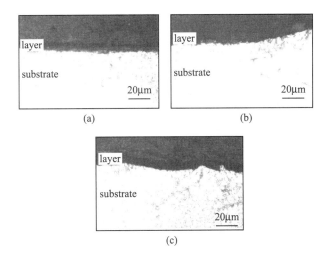

(a)   (b)

(c)

Fig.3.8   Cross-section morphologies of (a) 1045 steel treated by sulfurizing and shot-peening with intensities of (b) 0.15mmA and (c) 0.25mmA + sulfurizing under light microscope

Table 3.4   Thicknesses of sulfide layers on 1045 steels and 52100 steels

Thickness of sulfide layer/μm	Sulfurizing	Shot-peening with 0.15mmA intensity + sulfurizing	Shot-peening with 0.25mmA intensity + sulfurizing
1045 steel	2—3	5—7	7—10
52100 steel	2—3	5—7	7—8

3.2.2.3   Morphologies and composition of sulfide layer

The surface morphologies of 1045 steel treated by sulfurizing and shot-peening with intensities of 0.15mmA and 0.25mmA + sulfurizing are shown in Figure 3.9. On the sulfide layer surface, the traces of shot-peening are still remained. The EDX analysis results are shown in Table 3.5. As the shot-peening intensity increased, the content of sulfur in the sulfide layer increased as well. Figure 3.10 shows the cross-section morphology of 1045 steel treated by shot-peening with 0.25mmA intensity. No transition layer existed between the sulfide layer and substrate.

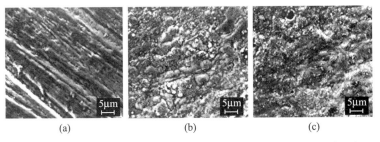

(a)   (b)   (c)

Fig.3.9   Surface morphologies of (a) 1045 steel treated by sulfurizing and by shot-peening with intensities of (b) 0.15mmA and (c) 0.25mmA + sulfurizing

Table 3.5   EDX analysis results of sulfide layers

Composition of sulfide layer	Atom%Fe	Atom%S
Sulfurizing	59.37	40.63
Shot-peening with 0.15mmA intensity + sulfurizing	55.22	44.78
Shot-peening with 0.25mmA intensity + sulfurizing	48.20	51.80

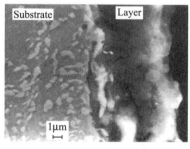

Fig.3.10   Cross-section morphology of 1045 steel treated by shot-peening with 0.25mmA intensity

3.2.2.4   Phase structure of sulfide layer

Figure 3.11 shows the phase structure of 52100 steel treated by sulfurizing and shot-peening with intensities of 0.2mmA and 0.4mmA + sulfurizing. The sulfide layer was composed of sulfide and substrate phases. The sulfide phases were FeS, FeS_2, and the substrate phases were α-Fe, Fe_5C_2.

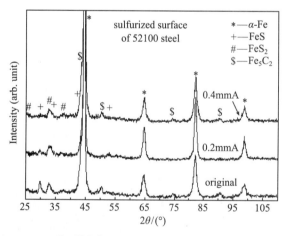

Fig.3.11   Phase structures of sulfide layers

3.2.3   Tribological Properties of Sulfide Layer

Friction and wear tests under dry condition were conducted on a SRV reciprocating tester. The upper samples were 52100 steel balls, 10mm in diameter with a hardness of HV 770. The lower samples were sulfurized and unsulfurized discs, 24mm in diameter with a thickness of 8mm. Experimental parameters: vibration amplitude was 1mm, frequencies was 15Hz, load was 50N, and experimental time was 2min. Friction and wear tests under oil lubrication condition were carried out on a ball-on-disc tester. The upper specimen was 52100 steel ball with 12mm in diameter and a hardness of HV 770. The lower samples were sulfurized and unsulfurized 1045 steel discs with 60mm in diameter and 5mm in thickness. The lubricant was engine oil with a dynamic viscosity of 37—43cSt at 50°C and the oil supplying rate was 1.9mL/min. The sliding speeds were 1.64m/s and 1.37m/s. The samples first passed a running-in period at loads of 1.2kg, 7.0kg and 12.8kg successively for 1min each. Then a load of 12.8kg was applied for 2min. The next load increment was 5.8kg. The running time per increment was 2min in a stepwise loading process until scuffing occurred. The wear-resistance experiments were conducted at the sliding speed of 1.15m/s. The samples first ran-in at the load of 7.0kg and 12.8kg for 1min successively, and then the load was up to 24.4kg. The experimental time was 18min. After experiments, the optical microscopy was used to measure the widths of wear scars of discs and balls.

3.2.3.1   Tribological properties of sulfide layer under dry condition

The values of friction coefficient μ_b at initial stage, the endurance life t_c (when $\mu = 0.3$) of the layer, and the μ_t at final stage are listed in Table 3.6. The shot-peening treatment did not extend the endurance lives t_c of sulfide layer. The friction coefficients μ_b was independent of the substrate surface state, layer thickness and substrate type. They were all between 0.15 and 0.18. The shot-peening had a little effect on the final friction coefficients μ_t, but the μ_t of sulfide layer on 1045 steel was slightly lower than that on 52100 steel. Figure 3.12 shows the comparison of wear scar widths. The wear scar widths of sulfurized steels were all smaller than those of unsulfurized steels, and shot-peening treatment could also decrease the wear scar widths. For the surface of shot-peening + sulfurizing steel, the wear scar width was smaller than that of steel treated by shot-peening or sulfurizing alone. The relative wear-resistance ε was regarded as the indexes to evaluate the wear-resistance of sulfide layer. ε was the ratio of wear scar width of sample untreated to that of treated, namely, $\varepsilon_S = w_{original} / w_{sulfurized}$, $\varepsilon_{P15} = w_{original} / w_{0.15mmA\ shot-peening}$, $\varepsilon_{P25} = w_{original} / w_{0.25mmA\ shot-peening}$, $\varepsilon_{P15S} = w_{original} / w_{0.15mmA\ combined\ treatment}$, $\varepsilon_{P25S} = w_{original} / w_{0.25mmA\ combined\ treatment}$, the values are listed in Table 3.7. It can be found that both shot-peening and sulfurizing increased the wear-resistance of the steel surface. The wear-resistance was further improved after shot-peening + sulfurizing combined treatment. Meanwhile, the wear scar widths of 52100 steels were lower than those of 1045 steels with the same surface states.

Table 3.6   Tribological behaviors of sulfide layers under dry condition

		μ_b	t_c	μ_t
	Sulfurizing		12	0.48—0.54
1045 steel	Shot-peening with 0.15mmA intensity + sulfurizing		8	0.48—0.57
	Shot-peening with 0.25mmA intensity + sulfurizing		6	0.50—0.56
	Sulfurizing	0.15—0.18	18	0.52—0.57
52100 steel	Shot-peening with 0.15mmA intensity + sulfurizing		6	0.48—0.60
	Shot-peening with 0.25mmA intensity + sulfurizing		9	0.54—0.59

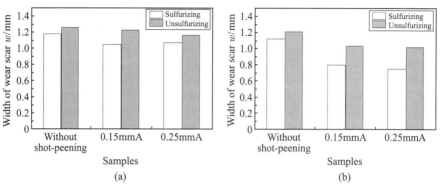

Fig.3.12   The wear scar widths of original steels and sulfide layers on (a) 1045 steel and (b) 52100 steel under dry condition

Table 3.7   Relative wear-resistance ε of 1045 and 52100 steels with different treatment

	ε_S	ε_{P15}	ε_{P25}	ε_{P15S}	ε_{P25S}
1045 steel	1.07	1.02	1.09	1.20	1.18
52100 steel	1.08	1.17	1.19	1.51	1.61

3.2.3.2   Tribological properties of sulfide layer under oil lubrication condition

Figure 3.13 shows the anti-scuffing loads P of samples with different treatment. The anti-scuffing loads of sulfurized surface were higher than those of unsulfurized surface, and the anti-scuffing loads of shot-peening surface increased slightly compared with those without shot-peening treatment. However, the anti-scuffing loads were independent of shot-peening intensity, indicating that the shot-peening treatment could not obviously improve the anti-scuffing properties. The variation of friction coefficient with load is shown in Figure 3.14. The fiction coefficients of sulfurized surfaces were lower than those of unsulfurized surfaces, but the shot-peening treatment had a little effect on the friction coefficients. Figure 3.15 shows the widths of wear scar of the discs and balls after the wear-resistance experiments. The shot-peening treatment

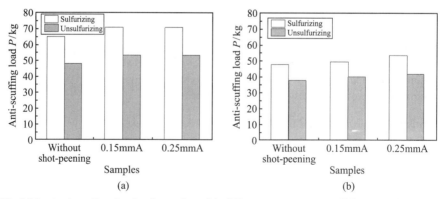

Fig.3.13   Anti-scuffing loads of samples with different treatment at a sliding velocity of (a) 1.37m/s and (b) 1.64m/s

could decrease the wear scar widths of the discs and balls, but they were independent of shot-peening intensity. The wear scar widths further decreased after shot-peening + sulfurizing combined treatment. The relative wear-resistance ε of the discs and balls were listed in Table 3.8. The discs and balls after combined treatment had the best wear-resistance. Meanwhile, the wear of the balls rubbed with the sulfurized discs was less than those of shot-peened discs, indicating that the sulfide layer played a good role in protecting its counterpart surface.

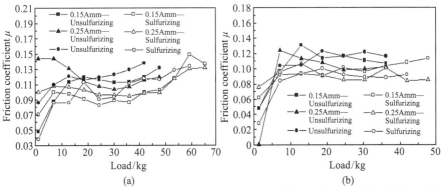

Fig.3.14 Variation of friction coefficient with load at a sliding velocity of (a) 1.37m/s (b) 1.64m/s

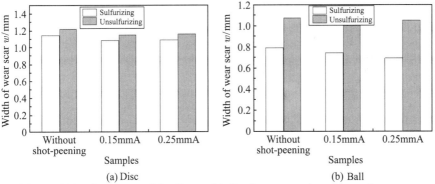

Fig.3.15 Widths of wear scar of the discs and balls after wear-resistance experiments

Table 3.8 Relative wear-resistance ε of the discs and balls with different treatment

	ε_S	ε_{P15}	ε_{P25}	ε_{P15S}	ε_{P25S}
Discs	1.07	1.06	1.05	1.16	1.12
Balls	1.35	1.07	1.02	1.45	1.55

3.3 Nitriding + Sulfurizing Combined Treatment

The sulfurized surface possesses good solid lubrication and anti-scuffing properties, however, its effect will be greatly reduced if the substrate is soft and can not give it an enough support. The nitriding layer is very hard and can significantly improve the wear-resistance and anti-scuffing properties of steel surface, but high hardness will aggravate the wear of counterpart surface and lead to high friction coefficient. According to the tribology principles, an ideal friction surface should be soft at the surface with good lubrication behavior, and the subsurface must be hard to give an effective support to the soft lubrication layer. Therefore, the combined treatment of nitriding + sulfuration is a reasonable design of friction surface.

3.3.1   1045 Steel Nitriding + Sulfurizing Combined Treatment

3.3.1.1   Preparation

Experimental materials were 1045 steel treated by quenching and high temperature tempering with a hardness of $HV_{0.05}$ 183 and roughness of $R_a = 0.8\mu m$. The samples were treated by nitriding at 560°C and sulfurizing at 200°C successively.

3.3.1.2   Structures

The phase structures, morphologies, and compositions of sulfide layers were analyzed by XRD and SEM + EDX.

Figure 3.16 shows the cross-section morphologies of nitrided layer and sulfide layer on 1045 steel under light microscope. After nitriding, a white light nitrided layer was formed evenly on the steel surface with a relatively high hardness about $HV_{0.05}$ 467. A sulfide layer was generated on the surface of the nitrided layer. Figure 3.17

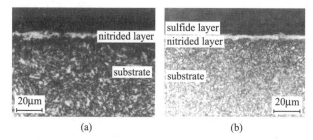

(a)   (b)

Fig.3.16 Cross-section morphologies of (a) nitrided layer and (b) sulfide layer under light microscope

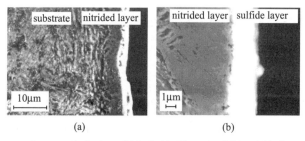

(a)   (b)

Fig.3.17 Cross-section morphologies of (a) nitrided layer and (b) sulfide layer by SEM

shows the cross-section morphologies of nitrided layer and sulfide layer under SEM.
Figure 3.18 shows the distributions of Fe, N and S elements along depth of substrates.
It was found that the thickness of the sulfide layer on the nitride layer was homoge-
neous, and combined closely with the latter.

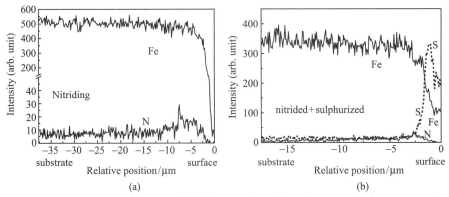

Fig.3.18   Distributions of Fe, N and S elements along the substrates and the cross-section
layers of (a) nitrided layer (b) nitrided layer and sulfide layer

Figure 3.19 shows the phase structures of the nitrided, sulfurized, and nitrided +
sulfurized layers. The nitrided layer was mainly composed of γ', Fe_3N, and ε-phase,

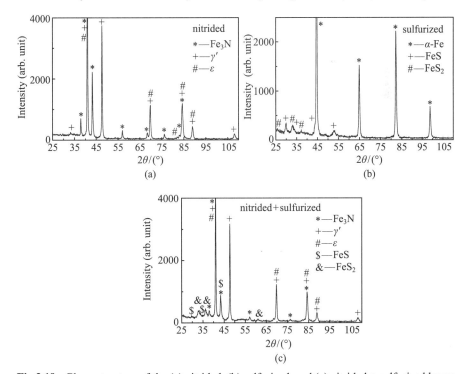

Fig.3.19   Phase structure of the (a) nitrided, (b) sulfurized, and (c) nitrided + sulfurized layers

while the sulfurized layer mainly included FeS, FeS_2, and α-Fe. After nitrided + sulfurized combined treatment, the α-Fe in the sulfurized layer was replaced by the phases in the nitrided layers, namely, the nitrided + sulfurized layers were composed by FeS, FeS_2, γ', Fe_3N, and ε-phase. However, the intensities of the nitrides in the composite layers were lower than those in the nitrided layer. This was that a part of nitride was decomposed during the sulfurizing process to produce iron atoms or ions which reacted with sulfur to generate iron sulfide.

3.3.1.3   Tribological properties

Friction and wear tests under dry condition were conducted on an SRV reciprocating tester. Experimental parameters: vibration amplitude was 1mm, frequencies was 15Hz, load was 50N, and experimental time were 30s and 2min.

Friction and wear tests under oil lubrication condition were carried out on a ball-on-disc tester. The sliding speed was 1.15m/s. The surfaces first ran-in at loads of 7.0kg and 12.8kg successively for 1min each, and then the load was up to 24.4kg. The experimental time was 18min.

1. Tribological properties under dry condition

The variation of friction coefficient with time is shown in Figure 2.35. The values of friction coefficient μ_b at initial stage, the endurance life t_c (when $\mu = 0.3$) of the layer, and the μ_t at final stage are listed in Table 3.9. w_{30s} and w_{2min} are the wear scar widths after 30s and 2min tests. The endurance life t_c of the nitrided + sulfurized surface was obviously longer than that of the sulfurized surface, and the t_c of the former was two times longer than that of the latter; meanwhile, the friction coefficient μ_b of the former was also slightly lower than that of the latter. The final friction coefficients μ_t were all lower than that of the plain surface, while the μ_t of the three treated surfaces was more or less constant and only the fluctuation of μ_t was relatively serious. The wear scar width of the plain surface was the biggest, and decreased after sulfurizing. The wear scar width of the nitrided surface was less than that of sulfurized surface and was only 40%—45% that of plain surface. The wear scar width of the combined treated surface was nearly the same as that of the nitrided surface,

Table 3.9   Tribological properties of all samples with different treatment under dry condition

	μ_b	t_c/s	μ_t	w_{30s}/mm	w_{2min}/mm
Plain surface	—	—	0.60—0.64	0.89	1.26
Sulfurized surface	0.15—0.17	7.3	0.48—0.58	0.64	1.18
Nitrided surface	—	—	0.55—0.60	0.40	0.53
Nitrided + sulfurized surface	0.14—0.15	15	0.54—0.59	0.41	0.54

2. Tribological properties under oil lubrication condition

Figure 3.20 shows the P-v diagrams of samples under oil lubrication condition. In the low speed range, both nitriding and sulfurizing could improve the anti-scuffing performance of the surface, while the anti-scuffing load of sulfurized surface was higher than that of nitriding. In the high speed condition, not either nitriding or sulfurizing was difficult to play the role of increasing scuffing load. However, the nitrided + sulfurized treatment could significantly improve the anti-scuffing performance of the surface at low and high speeds. The anti-scuffing load of the surface with combined treatment even exceeded the loading limit of the tester at low speed; it was about three times that of nitrided, sulfurized and plain surfaces. Figure 3.21 shows the variation of friction coefficient with load at sliding speeds of 1.10m/s, 1.37m/s and 1.64m/s. The friction coefficient of the plain surface was slightly higher, while there was no obvious difference among the nitrided, sulfurized, and nitrided + sulfurized surfaces. Figure 3.22 shows the wear scar widths of all discs and rubbing balls after wear experiments.

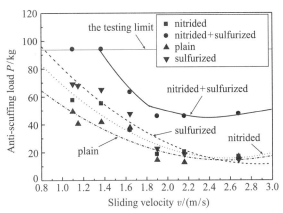

Fig.3.20 P-V diagram of all samples with different treatment

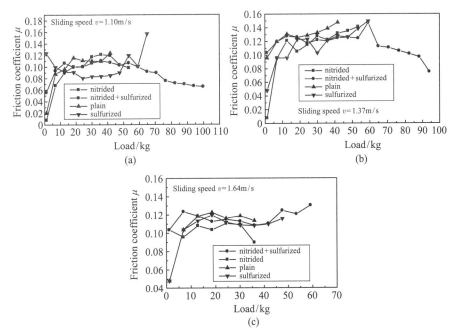

Fig.3.21   Variation of friction coefficient with load of the surfaces at sliding speeds of (a) 1.10m/s, (b) 1.37m/s and (c) 1.64m/s under oil lubrication condition

The wear scar widths of the modified surfaces were all less than those of the plain surface; meanwhile, the wear scar widths of balls rubbing with the nitrided and sulfurized discs were less than those rubbing with the plain surface. The sulfurizing treatment was more effective than nitrided treatment to protect the counterpart from wear. However, the wear scar width of the ball rubbing with the nitrided + sulfurized surface was more than that rubbing with the plain surface. Figure 3.23 shows the cross-section

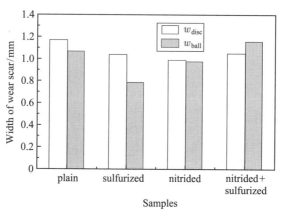

Fig.3.22   Wear scar widths of all discs and rubbing balls after wear experiments

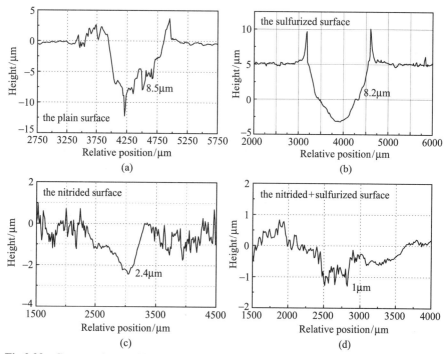

Fig.3.23   Cross-section profiles of wear scars on the discs, the marking data represent wear scar depths

profiles of wear scars discs, and the marking data in the figure represent the wear scar depths. It was found that the wear scar depth of the nitrided + sulfurized surface was very shallow no more than 1μm. The wear volume can be calculated by wear scar width and depth, shown in Table 3.10. The wear-resistance of the modified surfaces was all improved; especially the wear volume of the nitrided + sulfurized surface was the least.

Table 3.10 Wear scar widths, depths, and volumes of all discs (The perimeter of the wear scar center line is 97.34mm)

	Wear scar width w/mm	Wear scar depth d/μm	Wear scar volume V/mm^3
Plain surface	1.17	8.5	0.896
Sulfurized surface	1.04	8.2	0.750
Nitrided surface	0.99	2.4	0.224
Sulfurized + nitrided surface	1.05	1.0	0.097

3. Morphologies and compositions of the wear scar surfaces

Figure 3.24 shows the worn morphologies of all discs and balls. The wear of nitrided and nitrided + sulfurized discs were slight, the original machining traces were still visible, especially the nitrided + sulfurized disc, the sulfur distribution on the wear scar surface of it is shown in Figure 3.25, there were still a lot of sulfides remained in the wear trace. The wear scars of the balls rubbing with the discs treated by nitriding and nitriding + sulfurizing were minor than those of other two balls. The composition of wear scar surfaces of all samples is shown in Table 3.11. The contents of sulfur and oxygen on the worn nitrided + sulfurized disc surface were much higher than those on the sulfurized disc surface. There was a certain content of sulfur on the worn ball rubbing with the nitrided + sulfurized disc, and the content of oxygen was also high. The oxygen content on the worn steel ball rubbing with the nitrided disc was high, while that rubbing with the plain surface was low. Figure 3.26 shows the distributions of the elements on the wear scar surface of all discs along depth. The oxygen distribution depth on the wear scar increased in the order of original, sulfurized, nitrided, and nitrided + sulfurized discs, indicating that the oxide content in the boundary film was increased, the sulfur distribution depth of the former was at least 0.2μm, about 40 times that of the latter.

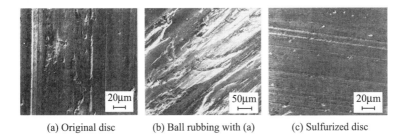

(a) Original disc   (b) Ball rubbing with (a)   (c) Sulfurized disc

(d) Ball rubbing with (c)　　(e) Nitrided disc　　(f) Ball rubbing with (e)

(g) nitrided+sulfurized disc　　(h) Ball rubbing with (g)

Fig.3.24　Worn morphologies of all discs and balls

Fig.3.25　Sulfur distribution on the wear scar of the nitrided + sulfurized disc

Table 3.11　Composition of wear scar surfaces of all samples

Samples	Atom%	Original surface	Sulfurized surface	Nitrided surface	Nitrided + sulfurized surface
Discs	Fe	92.88	98.51	91.34	85.19
	O	6.12	0.92	4.95	5.47
	S	—	0.57	—	5.36
	N	—	—	3.71	3.98
Balls	Fe	98.84	95.19	71.92	90.17
	O	0.06	3.72	26.45	7.93
	S	—	0.46	—	0.40
	Cr	1.10	0.63	1.62	1.50

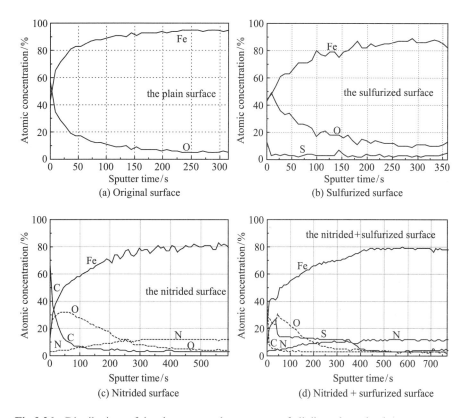

Fig.3.26   Distributions of the elements on the wear scar of all discs along depth

3.3.2   Gray Cast-iron Nitriding + Sulfurizing Combined Treatment

The nitrided gray cast-iron was treated by ion sulfuration, and its plain, nitrided, and nitrided + sulfurized surfaces were studied on a reciprocating sliding tester and a SRV friction and wear tester under dry and oil lubrication conditions [1].

3.3.2.1   Experimental methods

The experimental material was gray cast iron HT 200, the dimensions of samples were ϕ25mm $\times$10mm, the hardness was $HV_{0.5}$ 197. The samples were first treated by nitriding at 560°C and then sulfurized at 200°C. Determination of friction coefficients was carried out on the SRV friction and wear tester, the frequency f was 30Hz, time t was 8min, and the load P was 20N. The wear tests were performed on the reciprocating sliding tester.

The samples for wear tests were the plain, nitrided, and nitrided + sulfurized gray cast-irons, with dimension of ϕ25mm $\times$ 10mm. The counterpart was 52100 steel ball with diameter of 10mm and hardness of HRC 60. The experimental conditions were: the frequency of the reciprocating motion 120/min, time 30min, and loads 4141N, 9131N, 14121N, 19111N, 24101N.

3.3.2.2   Structures

Figure 3.27 (a) and (b) show the cross-section morphology and corresponding EDAX analysis of the nitrided + sulfurized composite layers. The white layer on the subsurface was nitride, while the sulfide layer was on the outer surface and presented dark gray color. The EDAX analysis indicated that there was high sulfur content on the sulfide layer.

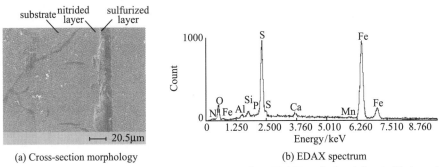

(a) Cross-section morphology   (b) EDAX spectrum

Fig.3.27   Cross-section morphology and corresponding EDAX analysis of the nitrided + sulfurized composite layer [1]

3.3.2.3   Tribological properties

1. Tribological properties under oil lubrication condition

Figure 3.28 shows the variation of the friction coefficient with time under oil lubrication condition. The friction coefficient of the plain surface was 0.160, and that of composite layer was 0.146 obviously lower than that of the plain surface.

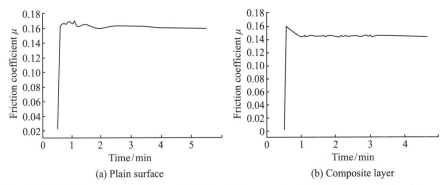

(a) Plain surface   (b) Composite layer

Fig.3.28   Variation of the friction coefficient with time under oil lubrication condition [1]

Figure 3.29 shows the variation of wear scar width with load under oil lubrication. The wear scar widths of the composite layer were always lower than those of the plain surface, while the wear scar widths of the nitrided layer were very close to those of the plain surface, and even wider at high loads (over 20N), which was caused by the high hardness of the nitrided layer. The wear scar width of the composite layer was the least due to the lubrication effect of sulfide layer.

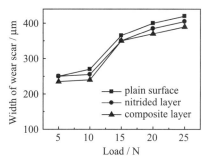

Fig.3.29   Comparison of the wear scar widths of all samples [1]

Figure 3.30 shows the worn morphologies (loading 19.11N) of the plain, nitrided, nitrided + sulfurized samples under oil lubrication condition.   There was a slight scratch on the worn surface of the plain gray cast-iron, while no scratch was observed on the worn nitrided + sulfurized surface.   Therefore, the combined treated sample exhibited excellent tribological properties.

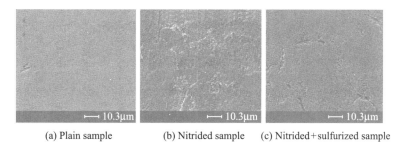

(a) Plain sample   (b) Nitrided sample   (c) Nitrided+sulfurized sample

Fig.3.30   Worn morphologies of all samples under oil lubrication condition

2. Tribological properties under dry friction

The variation of friction coefficient with time under dry friction is shown in Figure 3.31. The friction coefficient of the plain sample was 0.087; it sharply increased at first and then descended suddenly; after sliding about 1min, it gradually increased again and achieved a steady state. The friction coefficient of composite layer changed

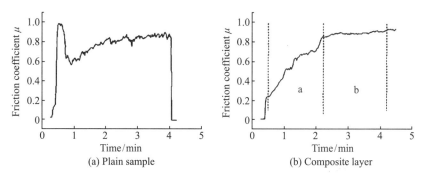

(a) Plain sample   (b) Composite layer

Fig.3.31   Variation of friction coefficient with time under dry friction condition [1]

in two regions: In region a, it was low due to the effect of sulfide layer and then gradually increased. In region b, it became stable with the value of 0.879 which was very close to that of the plain sample. The curve of the composite layer was more stable compared with that of the plain sample.

Figure 3.32 shows the worn morphologies (loading 19.11N) of the plain, nitrided, nitrided + sulfurized samples under dry condition. Obvious furrows appeared on the worn plain surface, there was visible fragmentation on the worn sulfide layer. A small number of scratches were on the worn surface of the composite layer, but obviously less than that on plain surface.

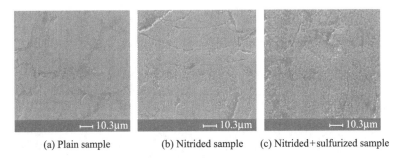

   (a) Plain sample   (b) Nitrided sample   (c) Nitrided+sulfurized sample

Fig.3.32   Worn morphologies of all samples under dry condition [1]

In brief, the nitrided and nitrided + sulfurized layers possess excellent tribological properties and can greatly improve the wear-resistance performance of gray cast-irons.

3.4   Nitrocarburizing + Sulfurizing Combined Treatment

CrMoCu alloy cast-iron is a material commonly used for engine cylinders. In order to increase the service lives of the engine cylinders, the surface of them can be ion nitrided to form a high hardness nitriding layer, which can significantly improve its wear-resistance and anti-scuffing properties, however, the increased hardness of cylinder surface will intensify the wear of the matched piston rings. An effective measure to lengthen the wear-life of the rubbing-pairs of cylinders and piston-rings is applying the combined treatment of nitrocarburizing + sulfurizing [2].

Figure 3.33 shows the variations of friction coefficient with time of the plain, sulfurized, nitrocarburized + sulfurized surfaces at the conditions of velocity 0.3m/s, load 80N, time 1667s, temperature 25°C, atmospheric environment and CD40 oil lubrication. The friction coefficient of the plain surface maintained at about 0.08 at the initial stage, after 1000s, the friction coefficient sharply rose with a piercing noise and the tester was forced to stop. The friction coefficient of the sulfurized surface was relatively low and kept at around 0.07 at the beginning of the test, after 1000s, the sulfide layer was damaged because of the low hardness of the substrate and the friction coefficient μ was increased as well. As a small amount of sulfide and renewable sulfide were still on the surface, the friction coefficient could remain at around 0.08. While the combined treated CrMoCu alloy cast-iron showed an excellent result, due to the effective support from the high hardness carbon-nitrogen compound layer on the subsurface, the sulfide layer was not easy to peel off. When the test was terminated, it still played the protective role.

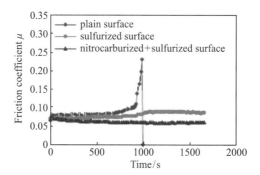

Fig.3.33 Variation of friction coefficient with time [2]

The combined treatment not only significantly enhanced the friction-reducing performance but also the wear-resistance of the CrMoCu alloy cast-iron, as shown in Figures 3.34 and 3.35. Under the same experimental conditions, the wear scar widths

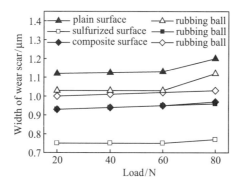

Fig.3.34 Variation of wear scar width with load [2]

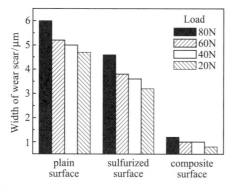

Fig.3.35 Variation of wear scar depth with load [2]

of the sulfurized surface and its matching ball were always the lowest at each load, while those of the nitrocarburized + sulfurized surface and its rubbing ball were higher, but the wear scar depths were obviously much lower than those of other counterparts, indicating that the wear-resistance of the CrMoCu alloy cast-iron was significantly improved after nitrocarburizing + sulfurizing combined treatment.

The combined treated CrMoCu alloy cast-iron also possessed very good friction-reducing property under liquid paraffin lubrication, as shown in Figure 3.36 [3]. The friction coefficient of the combined treated surface decreased by 16% and 81% respectively compared with that of the sulfurized and unsulfurized surfaces. The friction coefficient of the combined treated surface decreased by 29% under liquid paraffin with sulfur-containing additive lubrication condition, compared with that under liquid paraffin lubrication only. Above experimental results indicated that the sulfurized layer and nitrocarburized + sulfurized composite layer both played a significant role for improving the tribological properties, especially the friction-reducing performance was further enhanced under liquid paraffin with sulfur-containing additive lubrication condition due to a synergistic effect of sulfurized layer and sulfur-containing additive.

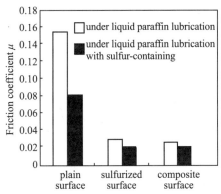

Fig.3.36   Friction coefficient under different lubrication conditions [3]

3.5   Thermal Spraying 3Cr13 Steel Coating + Sulfurizing Combined Treatment

3.5.1   Arc Spraying Technology

3.5.1.1   Principle

Arc spraying is a process of using the arc as heat source to atomize the melted wire by high speed airflow and spray on the workpiece surface to form a coating. Figure 3.37 shows the scheme of arc spraying.

Two filamentous wires are evenly and continuously fed into the two nozzles of the spraying gun by a wire feeding system. The nozzles are connected with the positive and negative electrodes of the power supply and the reliable isolation should be guaranteed before the contact of two wires. When the tips of two wires contact with each other, the arc is generated to melt the wire tips instantly; the melting metal is atomized to micro-droplets by the compressed air and then they are sputtered on the workpiece surface in a high speed to form arc spraying coating.

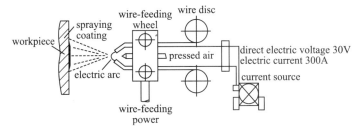

Fig.3.37   Scheme of arc spraying

3.5.1.2   Characteristics

(1) High bonding strength can be achieved without increasing the temperature of the workpiece and using the precious background materials. Sometimes aluminum bronze is used as the background material for arc spraying, the bonding strength of the spraying coating is greater than or equal to 20MPa. In general, the bonding strength of the arc spraying coating is 2.5 times higher than that of flame spraying coating.

(2) The weight of the spraying metal is large in unit time. The productivity of arc spraying is proportional to the arc current. When the spraying current is 300A, the productivities of spraying Zn, Al, stainless steel are 30kg/h, 10kg/h and 15kg/h, respectively. The productivity is increased by 2 to 6 times compared with that of flame spraying.

(3) The effect of energy-saving is very prominent and the energy utilization ratio is significantly higher than that of other spraying methods, as shown in Table 3.12. The energy costs decrease by more than 50%.

Table 3.12   Energy utilization ratios of several thermal spraying methods

Thermal spraying method	Energy utilization ratio/%
Arc spraying	57
Plasma spraying	12
Flame spraying	13

(4) The energy utilization ratio is high; meanwhile, the price of electric energy is far lower than that of oxygen and acetylene. Therefore, the cost is usually only 1/10 that of flame spraying; the investment in equipment is usually less than 1/3 that of plasma spraying.

(5) Arc spraying only uses compressed air and does not use combustible gases like oxygen, and acetylene. Therefore, the security is high.

3.5.1.3   Process

Arc spraying process includes surface preparation, spraying, and post-spraying treatment. A variety of spraying materials are heated to melting state and sprayed on the workpiece surface. In order to improve the bonding strength of spraying coating with substrate, the workpiece surface should be well pre-treated.

1. Surface cleaning

The spraying surface should be cleaned by degreasing, decontaminating, and derusting, etc.

2. Surface pre-processing

The pre-processing amount depends on the coating thickness of the design requirements. For the used workpieces, in general, they should be cut to below the largest amount of wear 0.1—0.25mm. For the new workpieces, the pre-processing amount of slip-fit surfaces should be 0.15—0.18mm or a bit larger.

3. Surface roughening

For hard working materials, sand blasting or sparkle galling can be used to roughen surface.

Sand blasting treatment can remove the oxide film on the workpiece surface and the "fresh" metal will be exposed to make the surface roughened. Meanwhile, a certain degree of compressive residual stress can be produced on the surface, which is beneficial to improve the fatigue strength after spraying.

After sand blasting, the spraying operation should be performed within 2—3h, otherwise, the surface can be contaminated, oxidized, and rusted again.

The choice of blasting materials can be determined by the size, shape and hardness of the workpiece. The efficiency is high by using corundum sand; the cast-steel shot has a big gravity and long service life, the sand grain size ranging between 18 mesh and 50 mesh. The coarse sand is used for strong and heavy workpieces. The blasting pressure ranges from 0.3MPa to 0.7MPa.

4. Standard parameters of arc spraying process

Table 3.13 shows the standard parameters of arc spraying process of ϕ3mm wires.

Table 3.13   Standard parameters of arc spraying process

Wire	Wire diameter/mm	Arc voltage/V	Current	Compressed air/MPa
Aluminum	3	34	150	>0.55
Zinc	3	28	120	>0.5
Aluminum bronze	3	35	200	>0.5
Carbon steel	3	35	200	>0.5

3.5.1.4   Effect of arc spraying process on the coating quality

1. Effect on the density of spraying coating

The density of the spraying coating depends on the size of melting metal particles at other constant conditions. The larger the particle size, the rougher the surface of the spraying coating. The processing factors affecting the density are mainly the pressure and flow of compressed air. In addition, the density is also related to the geometry of the nozzle of arc spraying gun, but has nothing to do with the wire diameter.

2. Effect on the hardness and bonding strength of the spraying coating

The hardness and bonding strength of the spraying coating mainly depend on following parameters.

(1) The chemical composition of spraying metal wire.

(2) The pressure of the compressed air.

(3) The volume of compressed air consumed by the spraying metal particles (tuyere diameter).

(4) The degree of pre-treatment of workpiece surface.

(5) The relative distance from the spray gun nozzle to the workpiece.

(6) The productivity of the spraying gun.

(7) The voltage on the electrode arc of the spraying gun.

With the increase of the pressure and flow (nozzle diameter) of compressed air, the hardness of the coating will be increased. Changing the productivity of the arc spraying gun and increasing the current can increase the temperature of metal particles before collision, namely the cooling duration is increased. As a result, the hardness of the coating will be decreased.

With the increase of compressed air pressure the impact force of metal particles is enhanced, which can increase the deformation of particles. Thus, the bonding strength of the coating is improved. When the air flow is increased, the same effect will be generated.

To increase the distance between the spraying gun nozzle and workpiece can decrease the jetting speed of metal particles and increase the oxidization degree. As a result, the bonding strength of the coating will be decreased.

In order to ensure the stability of spraying gun arc, to increase the voltage of electrode is necessary sometimes. However, the arc pressure should be minimized from the point of view of the metal loss, particle atomization quality, hardness and wear-resistance of spraying metals.

3.5.2   High-velocity Arc Spraying

High-velocity arc spraying is a new type of this technique. This spraying method has many advantages: The jetting speed of droplets is high; the atomization effect is good; the airflow speed can reach above 600m/s in the distance of 80mm away from the spraying gun; the bonding strength of the coating is significantly improved and the porosity of the coating is low.

In addition, high-velocity arc spraying also possesses a series of advantages such as low cost, high efficiency, convenient operation, and strong adaptability. Therefore, in recent years, it becomes the emphasis of propagation. Some researchers adopted the high-velocity arc spraying technology to prepare a 3Cr13 coating on the 1045 steel, and then treated it by sulfurizing to obtain the composite 3Cr13/FeS layer, which could significantly improve the wear-resistant properties of steels [4].

3.5.3   Preparation

The substrate material was 1045 steel with the hardness of HRC 55. The surface was grinded to an average surface roughness of R_a=0.8μm. The 3Cr13 coating with thickness of 300μm was first prepared by high velocity arc spraying using an HAS-01 gun. The wire composition is listed in Table 3.14. The spraying parameters were: The distance from nozzle to substrate 150mm, the spraying voltage 35V, the current 160A, and the pressure of the compressive air 0.7MPa.

Table 3.14   Wire composition for HAS-01 gun

Element	C	Si	Mn	Cr	S	P	Fe
Content/wt%	0.26—0.40	≤1.00	≤ 1.00	12—14	≤ 0.03	≤ 0.03	Bal.

The surface of the prepared 3Cr13 coating was grinded to a roughness of R_a=0.8μm and then was treated by sulfurizing for 2h to form FeS solid lubrication layer with a thickness of 3μm.

3.5.4   Microstructures

3.5.4.1   Surface morphology

Figure 3.38 (a) shows the surface morphology of the composite 3Cr13/FeS layer. It was found that the surface was relatively compact and smooth. Figure 3.38 (b) shows the cross-section morphology and element distribution of the composite 3Cr13/FeS layer. The bonding between the FeS layer and 3Cr13 coating was comparatively compact.

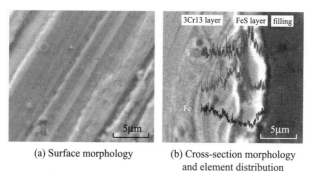

(a) Surface morphology   (b) Cross-section morphology
                                          and element distribution

Fig.3.38   Surface and cross-section morphologies and element distribution of the composite 3Cr13/FeS layer

3.5.4.2   Phase structures

Figure 3.39 shows the XRD pattern of the composite 3Cr13/FeS layer. The main phases were α-Fe and FeS.

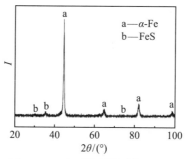

Fig.3.39   Phase structures of the composite 3Cr13/FeS layer

3.5.4.3 The residual stress

The residual stress was measured with an X-ray stressometer using $\sin^2 \Psi$ method. Figure 3.40 shows the plots of 2θ-$\sin^2 \Psi$ for the 3Cr13 coating and composite 3Cr13/FeS layer, θ represents the scanning angles; Ψ represents the side-tipping angles. The experimental results showed that the compressive residual stress on the surface of 3Cr13 coating was -164MPa, and the residual stress changed to -209MPa after sulfurizing. The compressive residual stress can prevent the initiation and propagation of cracks on the coating surface and extend its service life.

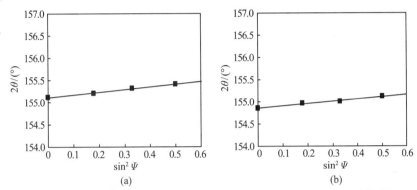

Fig.3.40 Plots of 2θ-$\sin^2 \Psi$ for the (a) 3Cr13 coating and (b) composite 3Cr13/FeS layer

3.5.4.4 Nano-mechanical properties

Figure 3.41 shows the curves of load-displacement for the 3Cr13 coating and composite 3Cr13/FeS layer. All the curves are nonlinear and smooth. This indicates that there are no cracks on the surface and the mechanical properties on different positions of the coating are homogeneous.

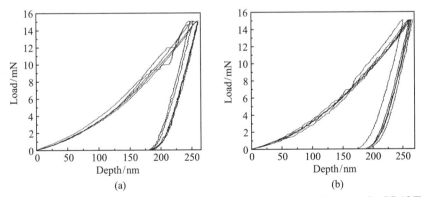

Fig.3.41 Curves of load-displacement for the (a) 3Cr13 coating and (b) composite 3Cr13/FeS layer

Table 3.15 shows the nano-hardness and elastic modulus of the composite 3Cr13/FeS layer.

Table 3.15   Nano-hardness and elastic modulus of composite 3Cr13/FeS layer

Sample	1045 steel	FeS film	3Cr13 coating	Composite 3Cr13/FeS coating
Nanohardness/GPa	10.51	2.84	10.66	10.17
Elastic modulus/GPa	247.76	52.97	226.40	230.32

3.5.5   Tribological Properties

The wear tests were carried out on a T-11 ball-on-disk tester under dry sliding and oil lubrication conditions. The upper samples were 52100 balls with 6.35mm diameter and HV 770 hardness. The lower samples were the plain, sulfurized, and sprayed + sulfurized 1045 steels with the diameter of 25.4mm and thickness of 6mm. Experimental parameters under dry friction condition were: load 5N, sliding speed 0.2m/s, and experiment time 60min. Experimental parameters under engine oil lubrication condition were: load 40N, sliding speed 0.2m/s, and experiment time 2h.

3.5.5.1   Tribological properties under dry condition

Figure 3.42 shows the curves of tribological properties of three samples under dry condition. Figure 3.42 (a) represents the variation of friction coefficient with time. At the beginning, the friction coefficient of composite 3Cr13/FeS layer was 0.026, and it gradually increased to a steady state of $\mu=0.124$. During the whole test, the friction coefficient of the composite 3Cr13/FeS layer was always lower than that of FeS film and 1045 steel. Figure 3.42 (b) represents the variation of wear scar depth with time. The wear scar depths of three samples all increased with time, but that of the composite 3Cr13/FeS layer was always obviously lower than others.

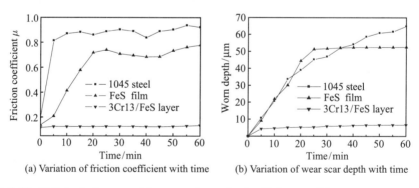

(a) Variation of friction coefficient with time   (b) Variation of wear scar depth with time

Fig.3.42   Tribological curves of three samples under dry friction condition

Figure 3.43 shows the worn morphologies and compositions of the 1045 steel, FeS film and composite 3Cr13/FeS layer under dry condition. The surface of 1045 steel was worn severely; there was a wide and deep worn scar. The worn extent of FeS film was slighter than that of 1045 steel. While the worn surface of composite 3Cr13/FeS layer was very smooth and only slight wear scar could be seen, a certain content of sulfur still existed on its worn surface after 60min sliding by the EDS analysis.

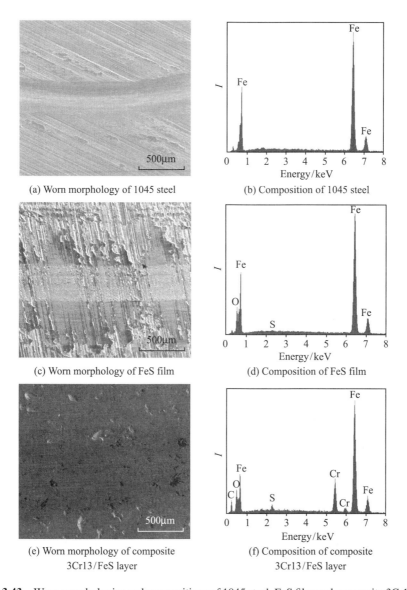

Fig.3.43 Worn morphologies and compositions of 1045 steel, FeS film and composite 3Cr13/FeS layer under dry condition

3.5.5.2 Tribological properties under oil lubrication condition

Figure 3.44 shows the tribological curves of composite 3Cr13/FeS layer and 1045 steel under oil lubrication condition. The starting friction coefficients of them were 0.1145, 0.111, respectively. At the first stage of sliding, the friction coefficient of them decreased gradually, and then reached to a stable stage; however, the friction coefficient of composite 3Cr13/FeS layer was always lower than that of 1045 steel. The wear scar depths of two samples increased with the increasing of time, but the wear scar depth of the composite 3Cr13/FeS layer was always lower than that of 1045 steel.

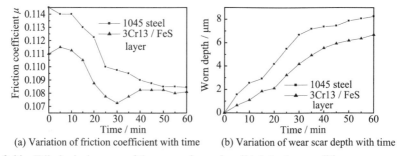

(a) Variation of friction coefficient with time   (b) Variation of wear scar depth with time

Fig.3.44   Tribological curves of three samples under oil lubrication condition

Figure 3.45 shows the worn morphologies and compositions of 1045 steel and composite 3Cr13/FeS layer under oil lubrication condition.

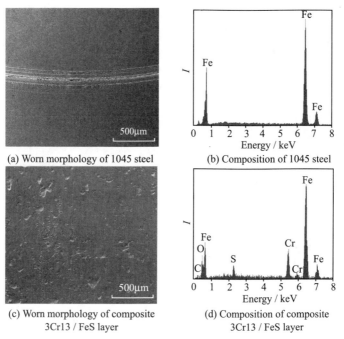

(a) Worn morphology of 1045 steel   (b) Composition of 1045 steel

(c) Worn morphology of composite   (d) Composition of composite
3Cr13 / FeS layer   3Cr13 / FeS layer

Fig.3.45   Worn morphologies and compositions of samples under dry sliding condition

3.6   Thermal Spraying FeCrBSi + Sulfurizing Combined Treatment

The velocity of the general plasma spray particles is about 200m/s; the velocity of the supersonic plasma spray particles can reach 400—600m/s. Therefore, the obtained coatings by supersonic plasma spray are more compact; meanwhile, their bonding strength with substrate is higher. A FeCrBSi coating on 1045 steel was prepared by the supersonic plasma spray technique, and then it was treated by sulfuration to form the FeS solid lubrication film. The composite FeCrBSi/FeS layer could significantly improve the tribological behaviors of steels.

3.6.1   Preparation

The substrate material was 1045 steel with the hardness of HRC 55 and surface roughness R_a=0.8μm. The composition of FeCrBSi self-fluxing alloy powder is shown in Table 3.16. FeCrBSi coating was prepared by supersonic plasma spraying. The substrate was sand blasted at first. The spraying parameters were: spraying voltage 140V, current 360A, power 40kW, and spraying distance 100mm. The FeCrBSi coating was then treated for 2h by low temperature ion sulfuration to obtain the FeS layer with thickness of 2μm.

Table 3.16   Composition of the FeCrBSi self-fluxing alloy powder

Element	Cr	B	Si	Mo	Fe
Content/wt%	13.8	0.55	0.51	1.81	Bal.

3.6.2   Microstructures

3.6.2.1   Surface morphologies

Figure 3.46 (a) shows the surface morphology of composite FeCrBSi/FeS layer. Its

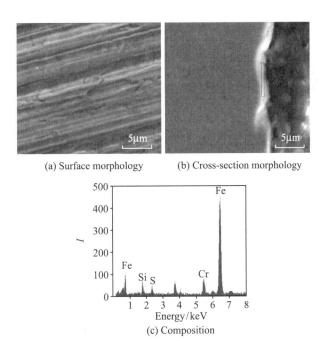

(a) Surface morphology   (b) Cross-section morphology

(c) Composition

Fig.3.46   Surface, cross-section morphologies and composition of the composite FeCrBSi/FeS layer

surface was comparatively compact with low porosity. Figure 3.46 (b) and (c) are its cross-section morphology and composition.

3.6.2.2   Phase structures

Figure 3.47 shows the phase structure of the composite FeCrBSi/FeS layer. It can be seen that the main phases are α-Fe and FeS.

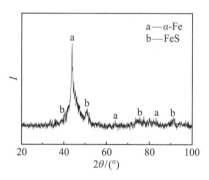

Fig.3.47   Phase structure of the composite FeCrBSi/FeS layer

3.6.2.3   Nano-mechanical properties

Figure 3.48 shows the curves of load-displacement for the FeCrBSi coating and composite FeCrBSi/FeS layer. The entire loading and unloading curves were smooth and continuous, indicating that no cracks appeared on the surfaces. The consistent curves in Figure 3.48 (a) demonstrated that the mechanical properties of FeCrBSi coating were homogeneous. However, the difference between the curves in Figure 3.48 (b) was probably due to the uneven distribution of the defects generated by ion

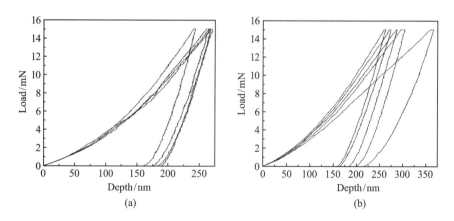

Fig.3.48   Curves of load-displacement for the (a) FeCrBSi coating and (b) composite FeCrBSi/FeS layer

bombardment during the sulfurizing process. Table 3.17 shows the nano-hardness and elastic modulus of different layers and the substrate.

Table 3.17   Nano-hardness and elastic modulus of different layers and substrate

Sample	1045 steel	FeS film	FeCrBSi coating	Composite FeCrBSi/FeS layer
Nanohardness/GPa	10.51	2.84	9.67	9.03
Elastic modulus/GPa	247.76	52.97	132.45	170.94

3.6.3   Tribological Properties

The friction and wear tester of model T-11 was used to investigate the tribological properties of the composite FeCrBSi/FeS layer in the atmosphere under dry and oil lubrication conditions. In addition, the tribological properties in air, vacuum, atomic oxygen radiation and ultraviolet radiation environments were investigated on a vacuum wear tester.

3.6.3.1   Tribological properties in the atmosphere under dry friction condition

Figure 3.49 shows the curves of tribological properties of three samples under dry condition. At first, the friction coefficients of 1045 steel and FeS film sharply increased and maintained at about 0.9, 0.7, respectively. While the friction coefficient of the composite FeCrBSi/FeS layer always kept at 0.1 and was very stable, indicating that the composite FeCrBSi/FeS layer possessed excellent friction-reducing performance. The wear scar depths of both 1045 steel and FeS film increased with time and achieved 60μm, 50μm, respectively after 1h. Nonetheless, the wear scar depth of the composite FeCrBSi/FeS layer was extremely low, no more than 2—3μm when the test terminated.

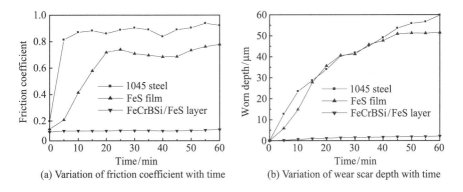

(a) Variation of friction coefficient with time   (b) Variation of wear scar depth with time

Fig.3.49   Tribological curves of three samples under dry friction condition

Figure 3.50 shows the worn morphologies and compositions of 1045 steel, FeS film and composite FeCrBSi/FeS layer after sliding 60min under dry sliding condition.

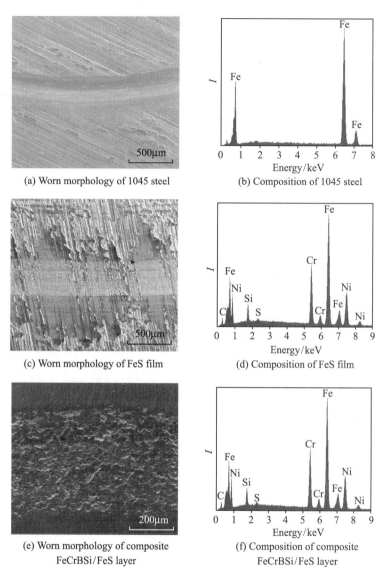

(a) Worn morphology of 1045 steel

(b) Composition of 1045 steel

(c) Worn morphology of FeS film

(d) Composition of FeS film

(e) Worn morphology of composite
FeCrBSi/FeS layer

(f) Composition of composite
FeCrBSi/FeS layer

Fig.3.50 Worn morphologies and compositions of the 1045 steel, FeS film and composite FeCrBSi/FeS layer under dry sliding condition

3.6.3.2 Tribological properties in the atmosphere under oil lubrication condition

Figure 3.51 shows the tribological property curves of three samples under oil lubrication condition.

Figure 3.52 shows the worn morphologies and compositions of 1045 steel and composite FeCrBSi/FeS layer after sliding 60min under oil lubrication condition.

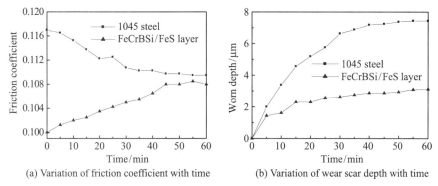

(a) Variation of friction coefficient with time   (b) Variation of wear scar depth with time

Fig.3.51   Tribological curves of the two samples under oil lubrication conditions

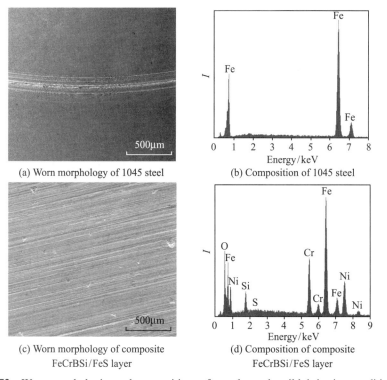

(a) Worn morphology of 1045 steel   (b) Composition of 1045 steel

(c) Worn morphology of composite   (d) Composition of composite
FeCrBSi/FeS layer   FeCrBSi/FeS layer

Fig.3.52   Worn morphologies and compositions of samples under oil lubrication condition

3.6.3.3   Tribological properties in air, vacuum, atomic oxygen radiation and ultraviolet radiation environments

1. Comparison of surface morphologies under different environments

Figure 3.53 shows SEM images of the composite FeCrBSi/FeS layer in air, vacuum, atomic oxygen radiation and ultraviolet radiation environments.   In air and vacuum environments, the composite coating is compact, and homogeneous. No obvious defect is observed. For the coating radiated by atomic oxygen radiation for 5h (see Figure 3.53 (c)), the surface is also compact. However, there are a few white particles with different sizes and homogeneous distribution. For the coating radiated by ultraviolet for 10h (see Figure 3.53 (d)), obvious pores and white particles are observed. The coating is relatively loose, but no obvious crack and collapse occurs. It can be concluded that atomic oxygen radiation and ultraviolet radiation have influence on the surface morphology of the composite FeCrBSi/FeS coating.

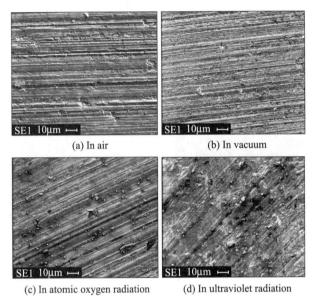

(a) In air   (b) In vacuum

(c) In atomic oxygen radiation   (d) In ultraviolet radiation

Fig.3.53 SEM images of the composite FeCrBSi/FeS layer in air, vacuum, atomic oxygen radiation and ultraviolet radiation environments

2. Comparison of friction coefficients under different environments

Figure 3.54 shows the variation of friction coefficients with time for original 1045 steel and the composite FeCrBSi/FeS layer in air, vacuum, atomic oxygen radiation and ultraviolet radiation environments under dry condition. The friction coefficients for the composite FeCrBSi/FeS coating in air and vacuum are almost identical, and keep at between 0.07 and 0.08. The friction coefficient for the composite FeCrBSi/FeS coating in ultraviolet radiation for 10h increases a little, and keeps at about 0.1. The increment of friction coefficient is probably attributed to the decomposition of FeS, leading to the decrease in lubricating property. The friction coefficient for the composite FeCrBSi/FeS coating in atomic oxygen radiation for 5h increases to 0.12—0.13. This is probably that part of the coating is oxidized, and the produced oxide decreases the lubricating property. In addition, the friction coefficients of the composite FeCrBSi/FeS coating in the above four environments are all lower than that of the original 1045 steel during the whole test. The composite FeCrBSi/FeS coating has good atomic oxygen radiation resistant and ultraviolet radiation resistant properties.

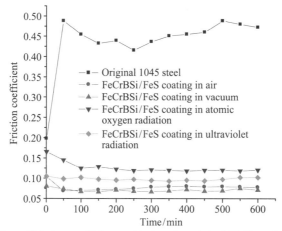

Fig.3.54   Variation of friction coefficients with time for original 1045 steel and the composite FeCrBSi/FeS layer in air, vacuum, atomic oxygen radiation and ultraviolet radiation environments

3. Comparison of worn morphologies under different environments

Figure 3.55 shows the worn morphologies of the composite FeCrBSi/FeS layer in air, vacuum, atomic oxygen radiation and ultraviolet radiation environments after worn 10min under dry condition. From Figure 3.55 (a) and (b), due to the self lubricating of FeS film, the worn track is shallow in air and vacuum environments. Slight adhesive wear and furrows occur. In these two environments, the samples operate smoothly with tiny noise. From Figure 3.55 (c) and (d), it can be seen that severe wear occurs when the composite coating is radiated by atomic oxygen and ultraviolet. Obvious wide and deep worn track accompanied with abrasive wear and adhesive wear is observed (see Figure 3.55 (e) and (f)). The width of worn track in atomic oxygen radiation and ultraviolet radiation environments is almost equal to that in air and vacuum, but a majority of FeS film has flaked off, and the wear loss increases. In atomic oxygen radiation and ultraviolet radiation environments, the samples operate unsmoothly with loud noise. It is concluded that atomic oxygen radiation and ultraviolet radiation decrease the lubricating property of the composite FeCrBSi/FeS layer.

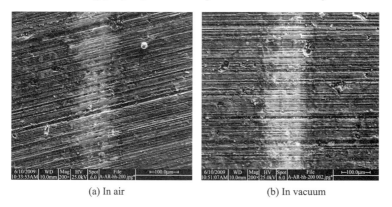

(a) In air   (b) In vacuum

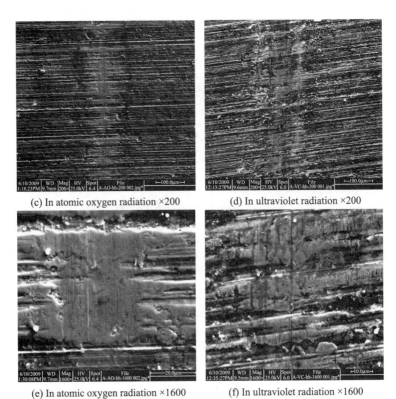

(c) In atomic oxygen radiation ×200   (d) In ultraviolet radiation ×200

(e) In atomic oxygen radiation ×1600   (f) In ultraviolet radiation ×1600

Fig.3.55 Worn morphologies of the composite FeCrBSi/FeS layer in air, vacuum, atomic oxygen radiation and ultraviolet radiation environments

Figure 3.56 shows the element distribution for the composite FeCrBSi/FeS layer radiated by atomic oxygen for 5h and ultraviolet for 10h. After 10min sliding, the sulfur intensity on the worn track declines, but sulfur and iron elements still exist, indicating that the atomic oxygen and ultraviolet radiation do not lead to the obvious change in the property of the composite FeCrBSi/FeS layer. Moreover, with the support of FeCrBSi layer, FeS film can continuously play the role of solid lubricating between the frictional pair.

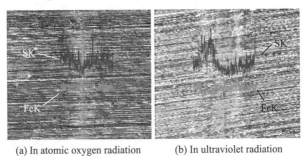

(a) In atomic oxygen radiation   (b) In ultraviolet radiation

Fig.3.56 Element distribution for the composite FeCrBSi/FeS layer in atomic oxygen radiation and ultraviolet radiation environments

4. Comparison of composition

Figure 3.57 and Table 3.18 are the composition and analysis results of energy spectra for the composite FeCrBSi/FeS layer in air and vacuum environments. From Figure 3.57, it can be seen that plenty of sulfur exists on the surface of the composite FeCrBSi/FeS layer. FeS sulfurized layer is obtained by chemical bond. During the sulfurizing course, the sulfur can diffuse into the substrate material through grain boundaries and dislocations and react with iron to generate FeS. The film is thin, but it combines well with the substrate. Moreover, the film surface is smooth and compact, which helps it play the anti-friction role continuously. During the friction process, FeS layer is rolled and adhere to the counterpart surface or fill into the valleys, which can prevent the direct contact between metals and the occurrence of adhesion and scuffing. However, with the deterioration of friction condition and prolonging of friction time, the friction heat is accumulated. When the temperature exceeds the melting point of FeS, it will be decomposed and loses the lubricating role. The micro-protrusions between metals will contact directly, the wear aggravates, and the friction coefficient increases. However, the role of friction heat is diplex, and it can make the decomposed sulfur atoms react with iron atoms to reproduce FeS. Thus, FeS film can play the lubricating role continuously.

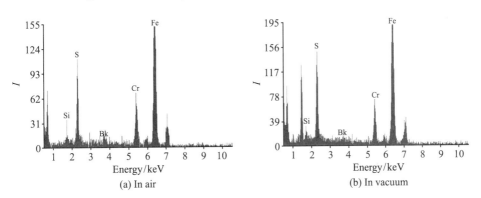

Fig.3.57   Composition of the composite FeCrBSi/FeS layer

Table 3.18   Analysis results of energy spectra (EDAX)

Elements	Fe	Cr	Si	B	S
In air/wt%	78.16	11.39	1.11	3.32	6.01
In vacuum/wt%	78.59	10.41	1.02	2.51	7.46

Figure 3.58 and Table 3.19 show the composition and analysis results of energy spectra for the composite FeCrBSi/FeS layer in a random area after atomic oxygen radiation. It is found that oxygen element occurs, indicating that the coating is oxidized. Figure 3.59 shows the composition of the white particle on the surface of composite FeCrBSi/FeS layer after atomic oxygen radiation. The content of oxygen increases, indicating that the white particle is iron oxide.

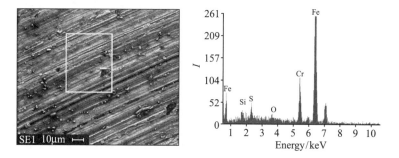

Fig.3.58   The composition of the composite FeCrBSi/FeS layer in a random area after atomic oxygen radiation

Table 3.19   Analysis results of energy spectra (EDAX)

Elements	Fe	Cr	Si	O	S
Particle/wt%	83.81	12.45	1.01	1.43	1.29
	83.68	11.67	0.95	2.13	1.57

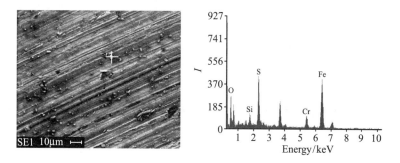

Fig.3.59   The composition of the white particle on the surface of composite FeCrBSi/FeS layer after atomic oxygen radiation

High energy atomic oxygen radiation has obvious influence on the tribological property, composition and surface morphology of the composite FeCrBSi/FeS layer. The grain boundaries of the composite layer contribute to the diffusion of high energy oxygen, leading to the oxidization of the layer. During friction, when the layer is subject to cyclic load, the oxide in the film is prone to fall off and form abrasive grains so that concave will occur, which will increase the friction coefficient and decrease the wearing life of the coating.

Figure 3.60 and Table 3.20 are the composition and analysis results of energy spectra for the composite FeCrBSi/FeS layer in a random area after ultraviolet radiation. Although some pores exist and coating gets loose, there is still sulfur on the coating surface, indicating that the FeS film can continuously play the role of lubricating role.

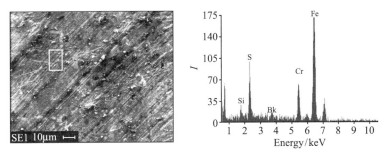

Fig.3.60   Energy spectrum of the composite FeCrBSi/FeS layer in a random area after ultraviolet radiation for 10h

Table 3.20   Analysis results of energy spectra (EDAX)

Element	Fe	Cr	Si	B	S
Ultraviolet radiation/wt%	78.78	11.78	1.14	3.27	5.04

Figure 3.61 and Table 3.21 are the composition and analysis results of energy spectra of the white particle on the surface of the composite FeCrBSi/FeS layer after ultraviolet radiation. The composition of the white particle is mainly Si, Fe and S. Ultraviolet radiation with high energy can make the chemical bond break, leading to the degradation of FeS. Ultraviolet radiation also can soften the coating surface and make it brittle so that cracks and collapse will occur, and finally decrease the mechanical property of the FeS layer.

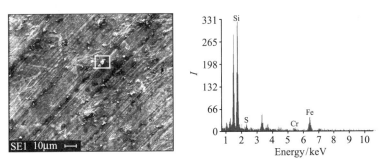

Fig.3.61   The composition of the white particle on the surface of composite FeCrBSi/FeS layer after ultraviolet radiation

Table 3.21   Analysis results of energy spectra (EDAX)

Element	Fe	Cr	Si	S
Ultraviolet radiation/wt%	38.94	4.62	53.18	3.26

3.7   MIG (metal inert-gas) Welding + Sulfurizing Combined Treatment

3.7.1   MIG Welding Technology

MIG welding is a special welding technique, it uses the metal wire equivalent to the parent material as the electrode, by the arc heat, the electrode is gradually melted and goes over into the molten pool in droplet. Its characteristic is using argon to make a thick and airtight gas-protective cover in the welding area so that the melting metal can avoid the oxidization by atmosphere [5].

Argon arc welding possesses many advantages as follows, compared with the manual arc welding:

(1) The argon used for arc welding ejects from the nozzle in laminar flow with a certain pressure and stiffness so that the protection is strong and stable.

(2) There is no intense chemical reaction: The argon is a single atom inert gas, it can not be decomposed at high temperature, does not react with melting metal, and is also not dissolved into the liquid metal. In the welding process, the molten pool is very quiet; no big spattering phenomenon occurs; the burning loss of the parent material is extremely less; the defects like gas cavity are not easy to be produced within the welding joint.

(3) The arc heat is centralized; the temperature of the arc column centre can achieve more than 10000K.

(4) There is no welding slag on the welding joint surface; it has a beautiful appearance.

(5) The heat affected zone is narrow; the weldment has a small deformation: As the arc column of argon arc welding is compressed, the heat is centralized, the temperature is high and the welding velocity is fast. Correspondingly, the heating degree of the parent material is low. Therefore, the deformation of the weldment caused by heating is smaller.

Some researchers used the MIG welding technique and wear-resistant welding wire UTP A DUR600 to prepare a surfacing layer on the 45CrNiMoVA steel substrate [6]. The structure of the welding joint was needle-like martensite. The average bonding strength of the layer was 695.3MPa. Its average micro-hardness was HRC 58.4. It showed a great improvement compared with that of the substrate, which was beneficial to increase the wear-resistance of materials.

3.7.2   Preparation

The researchers utilized the MIG welding technique to prepare a 500μm thick coating on the 1045 steel and then it was sulfurized.

The welding machine was a model WSE-500 with AS-DC pulse power supply and tungsten electrode. Its principle scheme is shown in Figure 3.62. The welding wire of UTP A DUR600 with ϕ1.2mm in diameter was used; its composition is listed in Table 3.22. The welding parameters were: current 85A, voltage 18V, velocity 5mm/s, protective gas argon (99.99% pure), and gas flow 8L/min. The surfacing layer was then sulfurized for 2h to form a 2μm thick FeS film.

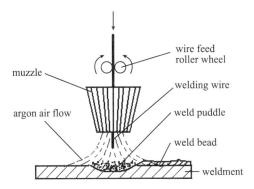

Fig.3.62 Scheme of the working principle of MIG welding

Table 3.22 Composition of the UTP A DUR600 welding wire

Element	C	Si	Mn	Cr	P	S	Cu	Fe
Content/wt%	0.47	3.10	0.40	9.15	0.023	0.002	0.19	Bal.

3.7.3 Structures

Figure 3.63 shows the surface, cross-section morphologies and EDX spectrum of the composite layer. It was found the surface was smooth and compact.

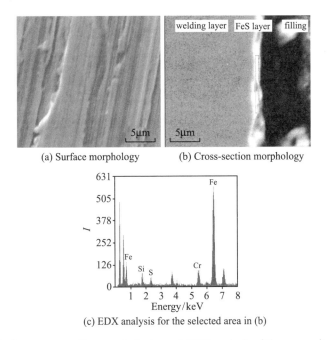

(a) Surface morphology   (b) Cross-section morphology

(c) EDX analysis for the selected area in (b)

Fig.3.63 Surface, cross-section morphologies and EDX analysis of the composite layer

Figure 3.64 shows the XRD pattern of the composite layer. The main phases were FeS, α-Fe and Fe—C solid solution.

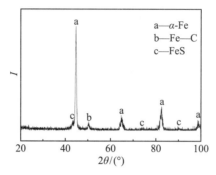

Fig.3.64　XRD pattern of the composite layer

Figure 3.65 shows the curves of load-displacement before and after sulfurizing. The five curves before sulfurizing show big differences, indicating that the mechanical properties of the welding layer were inhomogeneous; however, the consistency of the five curves after sulfurizing was improved. Table 3.23 shows the data of nano-mechanical properties of all samples.

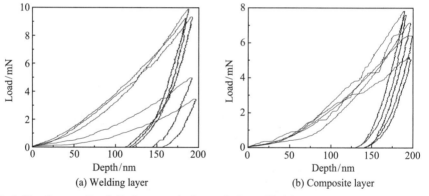

Fig.3.65　Curves of load-displacement before and after sulfurizing

Table 3.23　Nanohardness and elastic modulus of all samples

Sample	1045 steel	FeS film	Welding layer	Composite layer
Nanohardness/GPa	10.51	2.84	8.50	7.33
Elastic modulus/GPa	247.76	52.97	222.25	215.48

3.7.4　Tribological Properties

3.7.4.1　Tribological properties under dry friction condition

Figure 3.66 shows the curves of tribological properties of three samples under dry sliding condition. The friction of the composite layer had a big fluctuation in the first

25min; then it was inclined to be stable and was significantly lower than that of 1045 steel and FeS film. Meanwhile, the wear scar width of the composite layer was always lower than that of other samples.

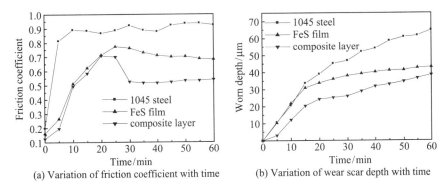

(a) Variation of friction coefficient with time

(b) Variation of wear scar depth with time

Fig.3.66   Tribological curves of three samples under dry friction condition

Figure 3.67 shows the worn morphologies and compositions of 1045 steel, FeS film and composite layer after sliding 60min under dry condition. The surface of the 1045 steel was worn seriously; the wear of FeS film was slighter than that of 1045 steel, while the wear of composite layer was the slightest, only a little coating was worn off, and more sulfur was still remained on the surface.

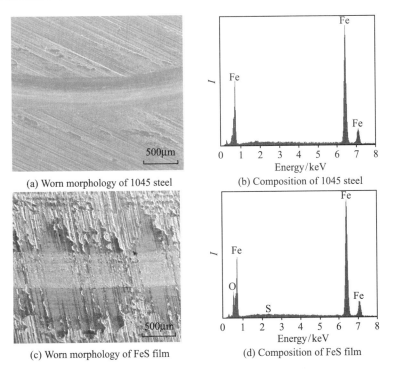

(a) Worn morphology of 1045 steel

(b) Composition of 1045 steel

(c) Worn morphology of FeS film

(d) Composition of FeS film

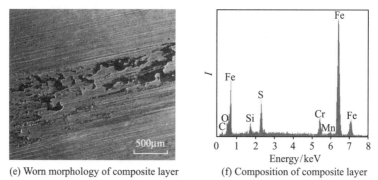

(e) Worn morphology of composite layer   (f) Composition of composite layer

Fig.3.67 Worn morphologies and compositions of 1045 steel, FeS film and composite layer under dry sliding condition

3.7.4.2 Tribological properties under oil lubrication condition

Figure 3.68 shows the curves of tribological properties of three samples under oil lubrication condition. The friction coefficient and wear scar depth of the composite layer were obviously less than that of 1045 steel and FeS film in the whole test.

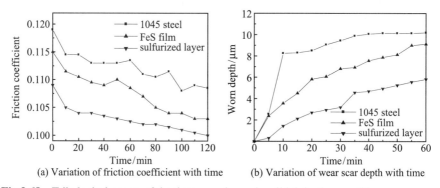

(a) Variation of friction coefficient with time   (b) Variation of wear scar depth with time

Fig.3.68 Tribological curves of the three samples under oil lubrication conditions

Figure 3.69 shows the worn morphologies and compositions of the 1045 steel, FeS

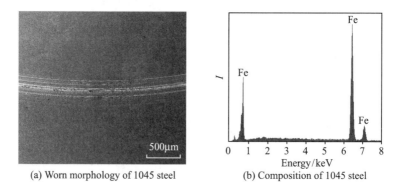

(a) Worn morphology of 1045 steel   (b) Composition of 1045 steel

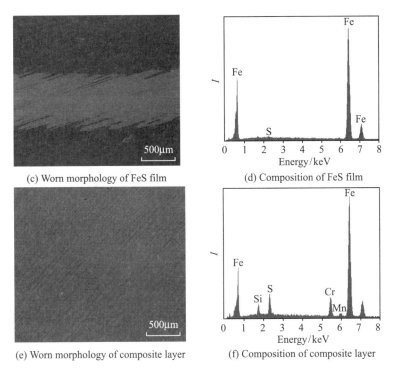

(c) Worn morphology of FeS film

(d) Composition of FeS film

(e) Worn morphology of composite layer

(f) Composition of composite layer

Fig.3.69   Worn morphologies and compositions of the 1045 steel, FeS film and composite layer under oil lubrication condition

film and composite layer after sliding 120min under oil lubrication conditions. Obvious wear scar had appeared on the surface of the 1045 steel; the surface of the FeS film had no visible wear scar, but the film had spalled off; the surface of the composite layer was still kept good.

3.8   Mechanism of FeS Film Prepared by Different Methods

(1) FeS film prepared by RF sputtering + sulfurizing.

Greater compressive stress was produced during the RF sputtering to increase the hardness of Fe coating, while FeS film was soft. As a result, the hard Fe coating gave an effective support to the soft FeS film, so that the tribological properties of steels were obviously improved.

During the friction process, the plastic flow of the soft layer took place. The surface roughness was decreased by the effect of peak load shifting. Therefore, the actual contact area became expanded and the contact stress descended. As a result, the friction coefficient was decreased. In addition, in the process of wear, the sulfide layer could continually play the role of solid lubrication by decomposition, diffusion, and regeneration. As the function of friction heat, the sulfide could be oxidized and precipitate active sulfur atoms. They would be partly oxidized and react with iron to form FeS again, so that the service life of sulfide layer could be extended. Meanwhile, the existence of iron oxide together with FeS could obviously increase the bearing ability of the steel surface.

(2) FeS film prepared by shot-peening + sulfurizing.

After shot-peening, the surface layer of the metal has greatly been plastic-deformed; its dislocation density increased; the microstructure was refined. As a result, the strength and hardness of the surface layer were improved. When the ions bombarded the metal surface in the sulfuration process, the newly formed and original defects in the material interacted with each other to increase the defect density of the substrate and promote the diffusion of sulfur, so that the sulfide layer could be thickened. With the increase of shot-peening intensity, the work-hardening phenomenon of the surface was more obvious and the depth of the sulfide layer was greater. The increased thickness of the sulfide layer extended the time of the diffusion of sulfur into the substrate; consequently, the sulfur was inclined to be gathered on the sulfide layer surface, leading to the increase of sulfur content on the surface. Finally, the FeS_2 phase increased, while the FeS phase decreased relatively.

Under dry friction condition, the friction coefficient of the combined treated surface was independent of the layer thickness, surface morphology and substrate state. The uneven surface morphology after shot-peening not only could not extend the endurance life of the sulfide layer but shorten it, which indicated that the rough surface could not enhance the attachment of the sulfide layer on the surface; conversely, it could easily lead to piercing the sulfide layer by the asperities and to the earlier direct contact between metals, which would accelerate the wear of sulfide layer. However, under the combined effect of sulfide layer and the increased strength and hardness by shot-peening, the wear-resistance of the surface was improved in a certain degree, compared with that after the single sulfurizing or shot-peening treatment.

Under oil lubrication condition, the boundary lubrication film composed of iron sulfide and iron oxide was formed on the frictional surface. With the increasing of load, the microstructures of the metal surface must be changed significantly under the friction heat and contact pressure. The cyclic softening phenomenon could happen on the shot-peened surface so that the dislocation density decreased, and compressive residual stress was gradually relaxed, which decreased the effect of hardening. Therefore, shot-peening could only slightly increase the anti-scuffing performance of the surface.

(3) FeS film prepared by nitriding + sulfurizing.

A large amount of nitrogen compounds like ε-phase and γ'-phase were produced on the iron surface after nitriding, it could significantly increase the hardness and red hardness of the surface layer, so that the anti-scuffing and wear-resistance performances were improved. During the friction, the sulfide layer prevented the direct contact between metals; meanwhile, as the function of contact stress and friction heat, the sulfide would be decomposed and then regenerated to form a boundary film composed of FeS and iron oxide on the surface. As a result, the anti-scuffing behavior was significantly improved. However, at high sliding velocities, with the increasing of load, more friction heat was generated, leading to severe wear of the sulfide layer. Therefore, the bearing abilities of the nitrided and sulfurized surfaces were lower. For the nitrided + sulfurized surface, the nitrided layer as the good support with high hardness and wear-resistance to the sulfide layer, the enduration life of sulfide layer was significantly extended, which avoided adhesion and reduced the friction heat, thus the fatigue wear of nitrided layer was also reduced.

(4) FeS film prepared by carbonitriding + sulfurizing.

As the sulfide layer on the outer surface was loose and porous, it was helpful to store oil. Meanwhile, The FeS was electrostatic polarized under the compressive stress to improve the bonding strength between the polar molecules in the oil and surfactant; as a result, the lubrication condition of the friction-pairs was significantly improved. In addition, the sulfide layer was very thin, and the actual contact area was small; thus, the thin and soft sulfide layer could effectively decrease the friction coefficient of friction-pair and reduce the tendencies of adhesion and scuffing. The soft sulfide could not only prevent the direct contact between metals to avoid or reduce the adhesion, it was also easy to transfer to its counterpart and continue to play the role of solid lubrication. Moreover, the carbon-nitrogen compound layer on the subsurface could give an effective support to the sulfide layer to avoid or reduce its spalling and extend its service life.

(5) FeS film prepared by spraying 3Cr13 + sulfurizing.

The content of chromium is relatively high (12%—14%) in the 3Cr13 coating, and chromium could react with iron to form Fe—Cr solid solution which could take the role of solution strength. During the spraying process, the hard phases of oxides of chromium were generated and distributed homogeneously, and they could take the dispersion-strengthened effect. The combined solution-strengthened and dispersion-strengthened effects can effectively mitigate abrasive wear.

The FeS layer was on the surface of the composite 3Cr13/FeS layer, with low hardness and shearing force. In the frictional process, it could prevent the direct contact between the metals to avoid the occurrence of adhesion. After a certain sliding time, the FeS layer was worn away gradually; however, the 3Cr13 coating could take the wear-resistant property continually. Therefore, the composite 3Cr13/FeS layer possessed excellent anti-wear and wear-resistance performances.

(6) FeS film prepared by spraying FeCrBSi + sulfurizing.

The content of chromium (13.8%) and boron (0.55%) was comparatively high in the FeCrBSi coating, and Fe—Cr solid solution could be generated and play the role of solution strengthening. Hard phases, like Cr_2B, Cr_2C_3 were generated and distributed homogeneously; they could play the role of dispersion strengthening. As a result, the combined solution and dispersion strengthening effects made the FeCrBSi coating possessing a high hardness (9.67GPa), and the coating could give a good support to FeS layer to improve the wear-resistant property. Furthermore, the FeCrBSi coating contained the molybdenum element, which could react with sulfur to form MoS_2 during the sulfurizing process. Therefore, the friction-reducing and wear-resistant properties could be greatly enhanced.

(7) FeS film prepared by Argon arc welding + sulfurizing.

The composite layer possessed excellent tribological properties due to the generation of the FeS film on the surface of welding layer. During sulfurizing, a large number of defects were produced on the welding layer surface by the action of ion bombardment, the produced dislocations and vacancies were beneficial to the diffusion of sulfur and generation of sulfide. In addition, the current density of MIG welding was high, the arc heat was centralized, the workpiece was subject to a little heat, and the cooling effect of inert gas flow could make the welding layer obtaining martensite microstructure with high hardness. Therefore, the welding layer prepared by argon arc welding technology had higher hardness and exhibited excellent wear-resistance.

References

1. Du M Y, Xiang N, Zhu Z X, et al. Tribological characteristics of the surface layer of grey cast iron modified by ion nitriding and ion sulfurizing [J]. Heat Treatment of Metals, 2005, 30(5): 12–15.
2. Hu C H, Ma S N, Li X, et al. Study on tribological properties of nitrocarburizing-sulphurizing layer of CrMoCu alloy cast iron [J]. China Surface Engineering, 2004, (4): 42–45.
3. Hu C H, Qiao Y L, Ma S N, et al. Study on friction and wear performance of nitrocarboni-ded-sulphurized layer of CrMoCu alloy cast iron under additive with sulphur lubricating [J]. Journal of Harbin Institute of Technology, 2006, 38: 176–179.
4. Kang J J, Wang C B, Wang H D, et al. Characterization and tribological properties of composite 3Cr13/FeS layer [J]. Surface & Coatings Technology, 2009, 203: 1927–1932.
5. Shan J G, Dong Z J, Xu B S. Development of surfacing technology and status of its application in basic industries [J]. China Surface Engineering, 2002, (4): 19–22.
6. Liu H B, Meng F J, Ba D M. Microstructure and properties of MIG welding surfacing layers on 45CrNiMoV steel [J]. China Surface Engineering, 2007, 20(3): 39–42.

Chapter 4
FeS Solid Lubrication Layer Prepared by Other Methods

The FeS solid lubrication layer prepared by low temperature ion sulfuration technique has been described in Chapters 2 and 3. Although its tribological properties are excellent, the applications are subject to a certain limitation due to the disadvantages of ion sulfuration technology. The costs of ion sulfuration are relatively high; meanwhile, the process must be completed in the vacuum chamber of a certain size, in other words, it can only deal with some smaller and easily transported parts, such as cylinder piston rings, nozzles, pistons, and roller bearings, etc. In order to extend the application of solid lubrication layer, it is very necessary to develop new processes for preparing low-cost FeS layers. In this chapter, other methods for preparing FeS coatings will be introduced.

Thermal spraying is a widely used technique in the surface engineering. It can prepare much thicker coatings and does not need to be carried out in vacuum chamber. Therefore, using thermal spraying technique to spray FeS powders on the surface of workpieces may become an easy, economical and practical means to prepare solid lubrication coatings.

Sol-gel is also a simple and effective technology to prepare solid lubrication coatings. It does not require special equipment. So this technique can be adopted to prepare various coatings for some heavy or complicated machine parts.

4.1　High-velocity Flame Sprayed FeS Coating

4.1.1　High-velocity Flame Spraying Technology

High-velocity flame spraying (HVOF) is a high-energy spraying technique appeared in the early 1980s. As the velocity of spraying particles is several times higher than that of powder flame spraying, it is also known as supersonic spraying.

4.1.1.1　The principle, spray gun construction and type of HVOF

For determinate materials, there are two approaches to improve the quality of spraying coatings [1]. One is to improve the heat flux of particles; the other one is to increase the speed of particles. The flame temperature of HVOF is generally around 2760°C, which mainly relies on the fuel and combustion chemical ratio. The heating of the particles are ensured by a high pressure in the combustion chamber and a long time staying in the nozzle, the spraying gun does not melt the powder particles but to improve the particle velocity to achieve supersonic speed with high kinetic energy. When bombarding the target, the impact of kinetic energy is converted into heat so that the particles are subject to sufficient deformation and a dense coating can be obtained.

In order to achieve the supersonic speed, the key point is to maximize the airflow speed in the nozzle. This mainly relies on the Laval nozzle designed by the principle of fluid dynamics. When the subsonic airflow enters the Laval tube, it becomes supersonic flow through the exit tube. It also uses the principle of detonation waves spreading in a supersonic speed, so that the explosion generates enough shock transmission and the chemical reactions are induced. Consequently, the combustible mixture in the shock tube explodes continuously to initiate detonation waves and form supersonic airflow.

The key structure of the spraying gun is an internal combustion chamber (Figure 4.1). The gas or liquid fuel is injected into the combustion chamber with the high pressure of 0.5—3.5MPa and high flow of 0.016m³/s. After the ignition, the continuous combustion and explosion take place to generate a high speed (4 to 5 times the speed of sound) and high temperature (2760°C) flame flow, which is jetted to the atmosphere through a long nozzle. Diamond speck in the flame flow can be seen in the atmosphere. Meanwhile, the powders are injected into the combustion chamber; they are heated in the flame flow to 1200°C and accelerated to 800m/s. Finally, the particles bombard the substrate to form a coating.

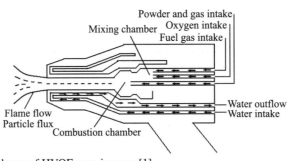

Fig.4.1   The scheme of HVOF spraying gun [1]

There are five kinds of typical HVOF spraying gun as follows.

(1) Spraying gun of model Jet Kote II made by Stellite Company (Figure 4.2).

The mixture of combustible gas and oxygen combusts in the combustion chamber located within the handle to generate high pressure and high speed hot air, which is divided into four jet strands through the 90° corner to enter the 150mm long Laval nozzle. The powders are injected into the nozzle by nitrogen from the central axis and are melted and speeded by flame flow. After compression and acceleration through the Laval nozzle, the particles are supersonic flow during leaving the nozzle. The combustible gas is propane.

(2) Spraying gun of model HV 2000 made by Miller Thermal Company (Figure 4.3).

The oxygen and combustible gas are injected into a mixing room and enter the combustion chamber to combust by lighting. Meanwhile, the powders are injected into the combustion chamber by gas flow from the central axis and are heated and speeded. The supersonic flow is formed after the compression by long nozzle. The combustible gases can be C_2H_2, H_2, C_3H_8, and natural gas, etc.

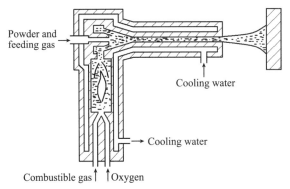

Powder and feeding gas

Cooling water

Cooling water

Combustible gas↑ ↑Oxygen

Fig.4.2   Spraying gun of model Jet Kote II [1]

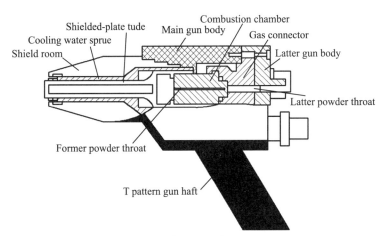

Shielded-plate tude   Main gun body   Combustion chamber
Cooling water sprue   Gas connector
Shield room   Latter gun body

                                                 Latter powder throat

Former powder throat

T pattern gun haft

Fig.4.3   Scheme of model HV 2000 spraying gun [1]

(3) Spraying gun of model CDS-100 made by Sulzer Company (Figure 4.4).

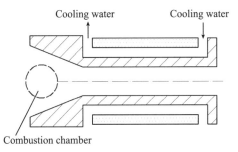

Cooling water   Cooling water

Combustion chamber

Fig.4.4   Scheme of model CDS-100 spraying gun [1]

The mixture of oxygen and combustible gases of C_3H_6, C_3H_8 or H_2 is injected into combustion chamber and electronic ignition is used to cause knocking. Meanwhile, the powder carrier gas flow is injected from the central axis. It is compressed and speeded when passing through the long shock tube to form supersonic flow.

(4) Spraying gun of model Diamond-Jet made by Metco-Perkin Company (Figure 4.5).

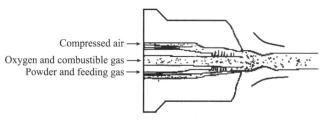

Compressed air →

Oxygen and combustible gas →
Powder and feeding gas →

Fig.4.5   Structure of model Diamond-Jet spray gun [1]

The mixture of oxygen and combustible gas (C_3H_6, C_3H_8 or H_2) is sent to the axial closed chamber in the exit of the gun. Meanwhile, the powder carrier gas flow is injected from the central axis and ignited to combust. The hot air in a circle flow is.compressed by external compressive air and speeded to form supersonic flow. The current design includes a so-called "extended air cover of gas-water compound cooling" composed of a water-cooling extended nozzle and compressive air, the purpose is to strengthen the heating and speeding of the particles.

(5) Spraying gun of model JP-5000 made by Hobart Tafa Company (Figure 4.6).

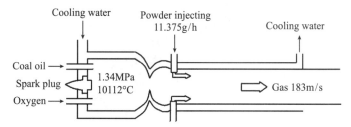

Cooling water   Powder injecting
                    11.375g/h   Cooling water

Coal oil →
Spark plug   1.34MPa
                    10112°C   Gas 183m/s
Oxygen →

Fig.4.6   Scheme of model JP-5000 spraying gun [1]

The fuel (liquid kerosene) and oxygen are injected into a combustion chamber behind a shrink-tensile nozzle from the central axis and ignited to combust generating a high pressure of 1.7MPa. Meanwhile, the powders are injected in a radial direction from the throat area (low pressure area) of the shrink-tensile nozzle and heated, speeded in the nozzle. The beam speed can achieve 100—1000m/s during leaving the nozzle. The generated coating is in a compressive stress state instead of normal thermal stress state.

4.1.1.2   The characteristics of HVOF process and coating performance

Compared with conventional spraying technologies, the main process characteristics are as follows [1].

(1) Compared with detonation gun spraying, the speed of combustion flame flow is 4 to 5 times the speed of sound. The particles have high speed and heavy impact on the substrate. The deformation of powder particles is sufficient to improve the bonding strength between the coating and substrate. Meanwhile, the porosity of deposited particles is small and the bonding between the particles is great.

(2) The air involved in the burning flame flow is small; meanwhile, the speed of the particles is high. Therefore, less spraying materials are oxidized.

(3) Burning flame flow is long and thin, with intensive energy. The distance between nozzle and workpiece can be increased; meanwhile, the operation is flexible. The distance between nozzle and workpiece of plasma spraying is generally 8—13cm, while the operation distance of HVOF can be 14—45cm.

(4) The ignition combustion, gas control and powder feeding of HVOF spraying equipment are all simple and easy to operate. The guide fire composed of hydrogen and oxygen can be ignited manually or automatically, meanwhile, the flow of combustible gas and oxygen is controlled by flowmeter. The flow of powder is controlled by electrical switch and can be adjustable automatically .

(5) Besides the ceramic powders with high melting points and powders with big particle size, the spraying powders can be all carbide and non-carbide powders. Meanwhile, for all plasma and D-gun formulations, the HVOF process has good potentials of application.

(6) The spraying efficiency is high. For many coating materials, the deposition efficiencies are all above 60%. The typical spraying velocity rate is 1—5kg/h and the coating with surface roughness of $R_a \geq 3\mu m$ can be achieved, shown in Table 4.1.

Table 4.1   Process comparison of HVOF and common spraying [1]

Process	Flame	Arc wire	plasma	Low-pressure plasma	detonation gun	HVOF
Heat source	$O_2, C_2H_2/$ H_2, O_2	Arc	Plasma arc	Plasma arc	$O_2,$ C_2H_2, N_2 explosion	Combustible gas, O_2
Promoting carrier	Air	Air	Inert gas	Inert gas	Explosion wave	Combustion ejection
Typical flame temperature/$^{\circ}C$	3000	5500	16000	16000	4500	3000
Typical particle speed/$(m \cdot s^{-1})$	30—120	240	120—600	<900	800	800
Average deposition speed/$(kg \cdot h^{-1})$	2—6	15	4—9	—	0.5	2—4
Coating porosity/%	10—20	—	2—5	<5	0.1—1	0.1—2
Mass fraction of oxide /%	4—6	0.5—3	0.5—1	10^{-6}	0.1	0.2
Relative bonding strength	General	Well	Well-excellent	Excellent	Excellent	Excellent
Relative technology cost (1 is the lowest)	3	1	5	10	10	5

As the particle speed generated by HVOF spraying is accelerated significantly compared with other spraying methods, the whole process of spraying can be improved. The advantages of obtained coating are shown in Table 4.2.

Table 4.2   Advantages and reasons of HVOF coating [1]

Coating advantage	Reason
Higher density (lower porosity)	Higher impact energy
Well corrosion-resistance	Less through-hole
Higher hardness	Less decomposition, higher bonding strength
Better wear-resistance	Harder, more resilient coating
Higher bonding strength	Improvement of bonding strength between particles
Less oxide content	Exposure to air in the short time
Less non-melting particles	Well heating particles
Powder chemical composition and phase component with strong stability	A short stay in the high temperature
Thicker coating (each and overall)	Improvement of residual stress
Smooth spraying surface	Higher impact energy

Table 4.3 shows the related performances of WC-Co coating by HVOF. Table 4.4 shows the characteristic comparison of HVOF sprayed and plasma sprayed Cu-Ni-In alloy coatings.

Table 4.3   Characteristics of coating with ω_{WC}=88% and ω_{Co}=12% [1]

Process	HVOF	D-gun	Air plasma	High-speed plasma
Flame temperature/° C	2760	2760	11100	11100
Flame flow speed	Mach4	Mach4	Subsonic	Mach1
Porosity/%	0	<1	<2	<1
Oxide mass fraction/%	<1	<1	<3	<1
Typical bonding strength/MPa	68.95	68.95	55.16	68.95
Maximum thickness/mm	1.524	0.762	0.635	0.381

Table 4.4   Characteristic comparison of HVOF sprayed and plasma sprayed Cu-Ni-In alloy coatings [1]

Performance	Normal bonding strength /MPa	Tangential bonding strength /MPa	Oxide mass fraction /%	Porosity /%	Micro-hardness /DPH300	Macro-hardness /HR15T	Deposition efficiency/%
HVOF	35.85	50.33	12	<0.5	230	89	75
PS	30.34	42.75	23	<0.5	160	83	45

4.1.1.3   Application and development of HVOF

High-velocity flame spraying was first used for aviation industry and turbine manu-facturing [2]. With the continuous development of application fields, the wear and corrosion problems have well been solved by this technique not only in the machin-ery manufacturing industry, but especially in various industries, such as paper, steel, polymer and power generation.

1. Steelmaking plant

The guide roller surface of galvanized iron sheet for car manufacture is worn seriously. In order to solve this problem, the HVOF coating technology was first developed and applied by Japanese. The results indicated that the HVOF coating improved the wear-resistance of the guide roller. The similar wear problem can be also encountered in the cold-rolling production.

2. Power plant

The boiler inner wall and combustion chamber are subject for long to the corrosion of carbon gray and high-temperature acid gas. In order to improve their corrosion-resistance, the dense corrosion-resistant protective coating can be obtained using ceramic and stainless coatings. Similarly, the anti-corrosion ability of heat exchangers can be also improved by this kind of coating.

3. Paper mill

In order to maintain normal production, to extend the service life of the machine, and to save resources, the coating can be used to repair the machine roller quickly and to strengthen the surface without disassembly of the parts.

HVOF coating technology has a wide range of application, especially has good economic benefits in the wear-resistance and corrosion-resistance aspects of ceramic coatings.

4.1.2   Preparation

The substrate material was AISI 1045 steel, treated by quenching and tempering to a hardness of HRC 55. The surface roughness was R_a=3.2μm. The FeS raw materials used for plasma spraying were block material with a purity of 85%, after ball-milling, FeS particles with 2μm in size were obtained. In order to avoid the burning of particles in the course of spraying, a layer of nickel-base alloy was coated on the external surface of FeS grains by chemical treatment. Three coatings with different thicknesses of 0.3mm, 0.8mm and 1.5mm were prepared for selecting an optimum thickness.

The FeS powders were sprayed using high velocity flame spraying equipment of model Jet kote. The mixture of O_2 and C_3H_4 was injected into the chamber and lighted by burning hydroxide flame. The gas with high temperature through the Laval nozzle achieved a supersonic speed. Meantime, the FeS powders were sprayed on the substrate surface by supersonic gas to form FeS coating.

4.1.3   Characterization

4.1.3.1   Morphologies of coating

Figure 4.7 shows the surface morphology of a 0.8mm thick plasma sprayed FeS coating by AFM, showing that the surface was formed by the accumulation of irregular FeS grains, and the surface was relative loose. It could be found that the grains' granularity was about 1μm, much less than that of the original grains (2μm); this phenomenon might be due to the burning and inter-pressing deformation in the course of spraying.

Figure 4.8 (a) and (b) are the surface morphologies of a 0.8mm thick FeS coating by SEM, which is nearly the same as that of other two thickness coatings. It was found that the FeS coating was formed by the pileup of many laminar units and was porous and incompact. Figure 4.8 (c) is the EDS analysis result from the same visual field of Figure 4.8 (a), showing a sharp peak of sulfur.

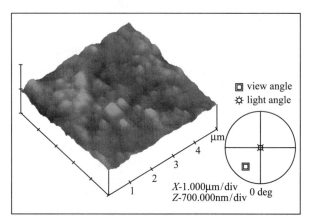

Fig.4.7   Surface morphology of FeS coating (AFM)

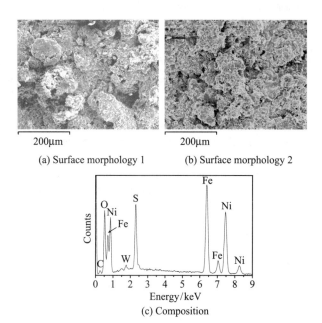

Fig.4.8   Surface morphologies and composition of the FeS coating (SEM+EDS)

Figure 4.9 shows the cross-sectional morphology of a 0.8mm thick FeS coating, which is nearly the same as that of other two thickness coatings. It can be seen that the coating was relatively compact near the interface, but was loose near the surface. A mechanical bonding between the coating and substrate was visible. However, the ion sulfurized coating showed a different characteristic, it was even and compact near the interface and surface. The transition between coating and substrate was also compact (see Figure 3.34).

800μm

Fig.4.9 Cross-section morphology of FeS coating

4.1.3.2 The phase structure

Figure 4.10 shows the X-ray diffraction pattern of a 0.8mm thick FeS coating, it was nearly the same as that of other two coatings. The phase structure of the FeS coating was quite complicated, showing that the oxidation and burning of FeS grains still occurred in the process of high temperatures during plasma spraying. The impurities might reduce the antifriction performance of coating.

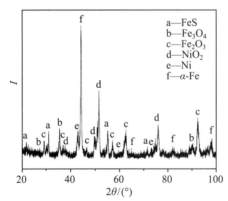

Fig.4.10 Phase structure of the coating

4.1.3.3 Bonding strength between coating and substrate

Figure 4.11 shows the bonding strength between the FeS coating and substrate. It can be seen that the bonding strength was 32N in average, does not change basically with coating thickness. The bonding strength between the FeS sprayed coating and substrate was much lower than that of ion sulfurized coating (49.3N).

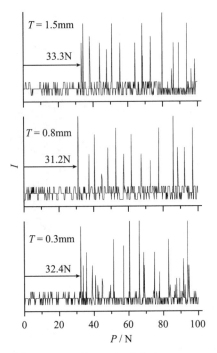

Fig.4.11   Bonding strength between coating and substrate

4.1.4   Tribological Properties of FeS Coating

The friction and wear tests under dry condition were carried out on a ring-on block tester of model MM-200. The upper samples were sulfurized 1045 steel blocks with dimensions of 31mm×6.5mm×6.5mm. The lower samples were quenched and tempered 1045 steel rings of 40mm in diameter and 10mm in width with a hardness of HRC 55. The experimental conditions were: room temperature, atmospheric environment, dry friction, sliding speed 0.42m/s and load 50N.

The friction and wear tests under oil lubrication were carried out on a ball-on-disc testing machine of model QP-100. The upper sample was a 52100 steel ball with a diameter of 12.7mm and hardness of HV 770, the lower samples were the sulfurized 1045 steel discs with dimension of ϕ60mm×5mm. All tests were carried out at room temperature, atmospheric environment (60%—70% humidity) and lubricated with No.40 base-oil (kinematic viscosity 37— 43mm^2/s at 50°C and velocity of oil supply 2mL/min) condition.

The wear loss was expressed by the width of wear scar. The scuffing load was identified by the weight when the coating initially scuffed. When testing the variation of friction coefficient with time, a load of 50N and velocity of 1.60m/s were fixed and the time changed from 0 to 60min. When testing the variation of wear scar widths with time, a load of 82N and velocity of 1.49m/s were fixed and the times were 7.5min, 15min, 22.5min, 30min and 37.5min. When testing the variation of scuffing load with velocity, the following velocities were employed 1.12m/s, 1.49m/s, 1.87m/s, 2.25m/s and 2.70m/s. The load was added stepwisely. The increment of load was 58N and the duration time was 2min. When the frictional force suddenly increased accompanying evident vibration and noise, this moment indicated the occurrence of scuffing, the sum of all weights was the scuffing load.

Figure 4.12 shows the variation curves of friction coefficient and wear scar width with time for plasma spraying FeS coatings and original 1045 steel under dry friction condition. It can be seen that the friction coefficient of each coating was stable with time after a running-in period. The friction coefficients of three coatings with different thickness were extremely close and obviously lower than that of original 1045 steel. The coating with 0.8mm thickness possessed the lowest friction coefficient. The wear scar width of each coating increased with time and thickness. When the thickness reached 1.5mm, the wear scar width of coating was even higher than that of original 1045 steel.

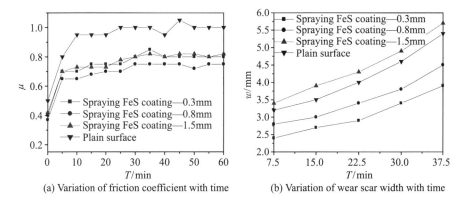

(a) Variation of friction coefficient with time   (b) Variation of wear scar width with time

Fig.4.12   Tribological curves under dry friction condition

Figure 4.13 shows the tribological curves of three coatings and original 1045 steel under oil lubrication condition. The variation of friction coefficient with time is showed in Figure 4.13 (a). It can be seen that the friction coefficient of each coating was stable and decreased slightly with time, the friction coefficient of FeS coatings was lower than that of the original 1045 steel; however, the thickness hardly influenced the friction coefficient. Anyway the coating with 0.8mm thickness showed the lowest friction coefficient.

Figure 4.13 (b) shows the variation of wear scar width with time. It can be found that the width increased with time and thickness, and the wear of 1.5mm thick coating was even higher than that of original 1045 steel, because the wear only took place within the coating.

Figure 4.13 (c) shows the variation of scuffing load with velocity. It can be found that the scuffing loads of three FeS coatings were all higher than that of 1045 steel. Along with the increase of velocity, scuffing load descended rapidly. In a word, with the increase of thickness, the anti-scuffing property of coatings was improved significantly, especially at low velocity range.

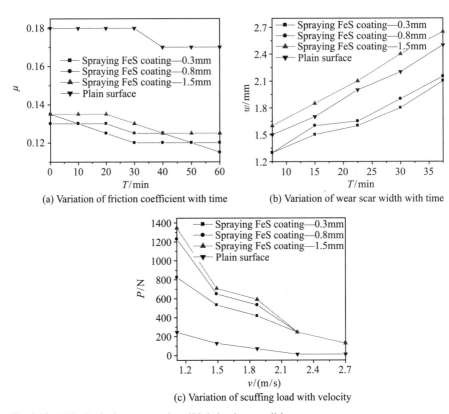

(a) Variation of friction coefficient with time   (b) Variation of wear scar width with time

(c) Variation of scuffing load with velocity

Fig.4.13 Tribological curves under oil lubrication condition

Figure 4.14 shows the variation curves of friction coefficient with load, under fixed velocity of 1.12m/s and 1.49m/s, respectively. The different length of curves means that they have different scuffing loads. It can be found that the friction coefficient increased with velocity, indicating that the coating might be destroyed greatly at a higher velocity. The friction coefficient was increased with load, whereas the sprayed coatings possessed lower value compared with 1045 steel and the 0.8mm thick coating had the best behavior.

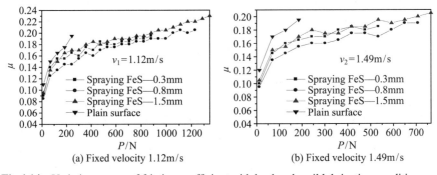

(a) Fixed velocity 1.12m/s   (b) Fixed velocity 1.49m/s

Fig.4.14 Variation curves of friction coefficient with load under oil lubrication condition

To sum up, the sprayed FeS coatings had much better friction-reduction, anti-scuffing, and wear-resistance properties than original 1045 steel, especially, the 0.8mm thick coating possessed the best tribological properties.

Figure 4.15 (a) is the wear scar morphology of 0.8mm thick FeS coating before wear test; it was just pressed by the vertical load (50N) from the opposite ring. At this moment, collapse has happened owing to the loose structure. Figure 4.15 (b) is the worn morphology after sliding for 1min; the FeS coating had become compact.

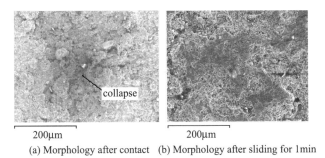

(a) Morphology after contact   (b) Morphology after sliding for 1min

Fig.4.15   Worn morphologies of the coating

Figure 4.16 (a) and (b) are SEM morphologies of worn surface of 0.8mm thick coa-ting, after sliding for 15min and 37.5min, respectively. The worn surfaces were

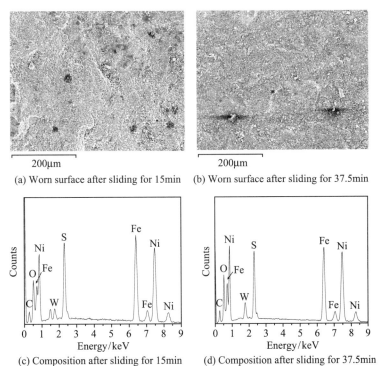

(a) Worn surface after sliding for 15min   (b) Worn surface after sliding for 37.5min

(c) Composition after sliding for 15min   (d) Composition after sliding for 37.5min

Fig.4.16   Worn surface and composition of the FeS coatings (SEM+EDX)

smooth and compact, compared with the original surface (see Figure 4.8 (a)) and the worn surface after sliding for 1min (see Figure 4.15 (b)). Figure 4.16 (c) and (d) are the EDS analysis results of the same visual fields of Figure 4.16 (a) and (b). It can be seen that they showed quite similar spectra, which means, the FeS coating was not worn through and the wear only took place within the coating.

Figure 4.17 is the scuffing morphology of a 0.8mm thick FeS coating, its failure mechanism is spalling. As discussed above, the bonding between the coating and substrate is mechanical, so the bonding strength is relatively low. During the scuffing experiment, once the shearing force was higher than the bonding strength the coating would be broken and flaked off.

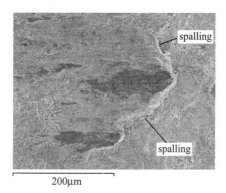

Fig.4.17   Spalling morphology of the worn FeS coating

Figure 4.18 (a) and (c) are the worn debris morphologies of a 0.8mm thick sprayed FeS coating and ion sulfurized coating after scuffing. It can be seen that the failure mode of the former was peeling off layer by layer; that of the latter was flaking. However, the worn debris size of the latter was much smaller than that of the former. Figure 4.18 (b) is the EDX analysis results for Figure 4.18 (a), showing the high content of sulfur. Furthermore, the distribution map of sulfur and other elements was very similar to that of the original sprayed FeS coating surface (see Figure 4.8 (c)). Figure 4.18 (d) is the EDX analysis results for Figure 4.18 (c), showing the low content of sulfur and high content of iron. That means, the ion sulfurized coating had been worn away mostly.

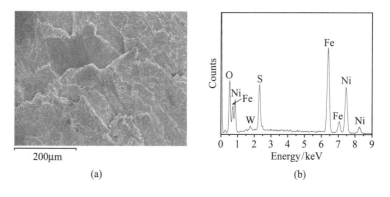

(a)   (b)

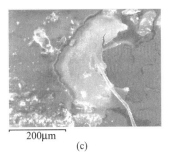
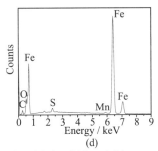

(c)   (d)

Fig.4.18 Morphologies and compositions of abrasive debris of (a) and (b) sprayed FeS coating, (c) and (d) ion sulfurized coating

4.1.5 Lubrication Mechanism of Sprayed FeS Coating

The forming process of a sprayed FeS coating is a process of continuous colliding, deforming and piling up of the melted or half-melted FeS particles. Therefore, the coating does not form a metallurgical but mechanical bonding with substrate. Its bonding strength is low (32.3N). However, when the plasma sprayed FeS coating comes to contact with the 52100 ball, the coating can prevent the direct contact between the metals and avoid the occurrence of adhesion. Some FeS particles can be transferred from the coating to the surface of 52100 steel ball, showing better tribological performance.

The coating with different thickness usually displays different effect. The thin coating possesses relatively good wear-resistance, whereas the thick coating has a relatively better anti-scuffing property. However, no matter how thin or thick, all coatings possess almost the same friction–reduction properties.

Due to the formation characteristics, the HVOF sprayed FeS coating contains more impurities, and porosities, and the thicker the coating, the more cavities it contains. In the wear test, under the vertical pressure from the steel ball, "collapse" or "spalling" of the coating are easy to occur, giving wider wear scar tracks. The wear scar width of a 1.5mm thick coating is even larger than that of substrate, because the coating is too thick, the wear happens only within the coating. But the collapse will stop when the coating becomes compact along with the wear process. So no matter how thick the coating is, the coatings with different thickness show similar friction coefficient.

During the scuffing experiment, the vertical pressure of the coating is increased with load and velocity, producing much frictional heat, the spalling and decomposition of FeS coating become even more severe. When the coating is worn out ultimately, the thicker the coating, the longer the service life it lasts, and the greater the load it can bear.

4.2   Plasma Sprayed FeS and FeS$_2$ Coatings

4.2.1 Plasma Spraying Technique

4.2.1.1 Overview of plasma spraying

The heat source of plasma spraying is compressive plasma arc [3]. When the arc is compressed in the plasma spraying gun, the energy is concentrated and the cross-section energy intensity can reach 10^5—10^6W/cm^2. The temperature in the center of arc column can be increased to 15000—33000K. In this case, the gas in the arc column will be ionized intensively and form a plasma body.

The characteristics of plasma arc are as follows.

1. High temperature and intensive energy

Figure 4.19 shows the measured temperature of a 400A non-transitive plasma arc with argon flow of 10L/min. The temperature in the centre of nozzle exit has achieved 20000K.

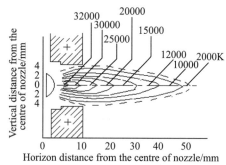

Fig.4.19   Temperature of a non-transitive plasma arc [3]

The high temperature and concentrated energy of plasma arc are of great practical value. In the process of spraying, welding and surfacing, the plasma arc can melt any kind of metals and non-metals. Meanwhile, a high productivity can be obtained and the deformation of weldment as well as welding heat affected zone can all be reduced. The plasma arc can also cut copper, aluminum, stain-less steels and titanium, which can not be cut by oxygen-acetylene flame.

2. High flame flow velocity

The working gas in the spraying gun is heated to ten thousands degree temperature, its volume expands fiercely. In consequence, the plasma flame flow is jetted with high speed and great impact force, which is beneficial to cutting and spraying processes.

Figure 4.20 shows the measured results of nitrogen plasma arc flame flow velocities for cutting (with gas flow of 40L/min, nozzle diameter of 2.8mm, current of 300A and voltage of 252V). It can be seen that the velocity of flame flow can achieve 2.4×10^3—4.2×10^3m/s at the distance of 10mm from nozzle tip. The plasma arc flame flow velocity for spraying is a little lower, usually a few hundred meters per second.

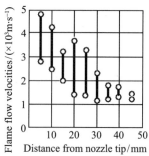

Fig.4.20   Velocities of plasma arc flame flow for cutting [3]

3. Good stability

As the plasma arc is a compressive arc, its arc column is tall and straight; its ionization degree is high. Therefore, the arc position and shape as well as arc current and voltage are all more stable than that of free arc, and they are less affected by the external factors. This is of great importance for ensuring the quality of spraying, welding, surfacing and cutting.

4. Good accommodation

There are more adjustable factors for the compressive arc. They can work stably in a wide range of parameters to meet the requirements of various plasma arc processes, which can not be achieved by free arc. For instance, oxidized, neutral or reductive gas atmospheres can be gained by changing the working gas type; rigidity-flexibility of plasma arc can be controlled by changing the nozzle size, gas flow and electrical parameters. In addition, for specific plasma arc equipment, flame flow temperature and jetting speed can be adjusted flexibly by regulation of electrical power to meet the requirement of different materials.

4.2.1.2   Principle and characteristics of plasma spraying

Figure 4.21 shows the schematic diagram of the principle of plasma spraying. The right part of the diagram is a plasma generator, which is also known as plasma spraying gun. Nitrogen or argon is introduced according to the requirement of technology. When these gases flow into the arc column region, they are ionized to become plasma. After the arc ignition, the arc is subject to three compressive effects in the pore channel and the temperature as well as spraying speed will be increased. At this moment, powder materials are transported to the powder feeding tube of the former gun. The powders are heated to the melting state in the plasma flame flow and are sprayed on the surface of workpiece. When the melting spherical powders impact on the substrate, the plastic deformation and adherence take place forming a spraying coating.

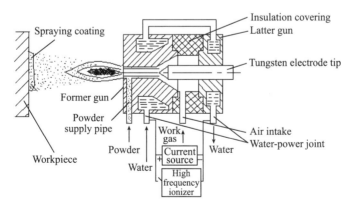

Fig.4.21   Schematic diagram of the plasma spraying [3]

The characteristics of plasma spraying are as follows.

(1) The parts do not deform and the original heat treatment state of substrate metal is not changed.

Although the temperature of plasma flame is quite high, the temperature of parts can be controlled within 200°C by a proper technology. This point is very beneficial to repairing the thin-wall parts, slender bar and some precise parts; some high strength steels can also be sprayed.

(2) More coating materials can be used by plasma spraying.

As the temperature of plasma flame is high, various materials can be heated to the melting state. Therefore, materials for the plasma spraying are extensive and coatings with a variety of performances can be prepared, such as wear-resistant coating, heat insulation coating, anti-oxidation coating, and electric insulation coating, etc. In this aspect, some other spraying techniques, like oxygen-acetylene flame spraying, arc spraying, high-frequency induction spraying and detonation gun spraying are all inferior to plasma spraying.

(3) The process is stable and the quality of coating is high.

The parameters of plasma spraying can be controlled quantitatively, therefore, the process is stable and the reproducibility of coating performance is great. During spraying, the deformation is sufficient when the melting particles collide with parts, thus the coating is dense and the bonding strength is high, its value can reach 40—70MPa, while that by oxygen-acetylene flame spraying is only 5—10MPa. The contents of oxygen and nitrogen in the coating can also be reduced significantly.

Table 4.5 shows the application and development of various thermal spraying methods.

Table 4.5   Application proportion of main thermal spraying methods

Spraying methods	1960	1980	2000
Wire fire spraying	35	11	4
Powder fire spraying	35	28	8
Wire arc spraying	15	6	15
Plasma spraying	15	55	48
High-velocity fire spraying	—	—	25

The disadvantages of plasma spraying are higher investment and necessity of pure nitrogen or argon gas. Meantime, the security measures should be strengthened for the health care of operators.

4.2.2 Preparation of FeS (FeS$_2$) Coating

Up to now, the spraying coatings are mostly used to improve the wear- and corrosion-resistance of materials. However, the application for improving the friction-reducing performance has not been developed. The spraying molybdenum coating used for friction-reduction purpose can be realized only by adding sulfur-containing additives in lubricating oil to generate MoS$_2$ lubricant on the sprayed Mo surface. But obviously, it is not applicable in dry friction condition. FeS is an effective solid lubricant with high melting point. The application of FeS coating must be very extensive if it is prepared by spraying method. As the spraying process does not need to be carried out in a vacuum furnace, so it is not limited by the furnace size; meanwhile, much thicker coatings can be fabricated. Therefore, spraying method is suitable for treating the friction pairs of large and heavy equipment.

Experimental material was AISI 1045 steel with a hardness of HB 162—165 and dimensions of 31mm×6.5mm×6.5mm. The FeS (FeS$_2$) coating was prepared by a spraying equipment of model Metco 9MB with GM-Fanuc 6-axis thermal spraying robot, and a 2-axis sample platform. The computer was used to control the spraying process. The spraying parameters are as follows.

Powder feed rate: 22.7g·min^{-1};

Current: 200—250A;

Voltage: 55V;

Spraying distance: 88.9mm.

4.2.3   Characterization of FeS (FeS$_2$) Coating

The cross-section morphologies and composition of FeS (FeS$_2$) coating were analyzed by SEM and EDS. The XRD was utilized to examine the phase structure.

Figure 4.22 shows the cross-section morphologies and distribution of Fe, O, and S elements of FeS (FeS$_2$) coating. The thickness of coatings ranged between 15μm and 20μm. The white particles in the pictures were sulfide which distributed dispersedly. The gaps between the particles were oxides. The hardness of coatings was ranging between HV$_{0.05}$130 and HV$_{0.05}$160. Figure 4.23 shows the phase structure of FeS and FeS$_2$ coatings, which were composed of α-Fe, FeS, FeS$_2$, Fe$_3$O$_4$ and Fe$_2$O$_3$, indicating that despite the protection of the Ar, FeS and FeS$_2$ were still oxidized in the process of spraying.

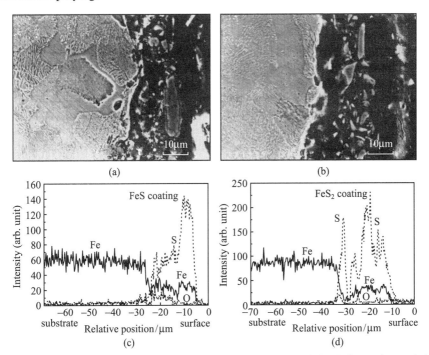

Fig.4.22   The cross-section morphologies of (a) FeS coating and (b) FeS$_2$ coating and the distribution of Fe, O, S elements

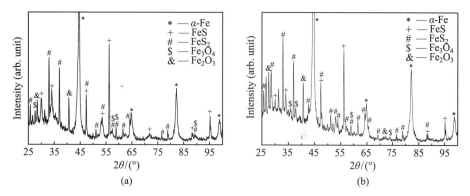

Fig.4.23   Phase structures of (a) FeS and (b) FeS$_2$ coating

4.2.4   Tribological Properties of FeS (FeS$_2$) Coating

The friction and wear teats were conducted on a block-on-ring tester of model MM-200. The upper specimens were 1045 steel blocks with and without sprayed coating. The lower samples were 52100 steel rings. The tests were carried out under dry friction condition. Experimental parameters were: rotational speed of lower sample 200r/min, load 5kg, 10kg, 15kg and 20kg, experimental time 7min, 15min, 22min, 30min, 38min and 45min. The worn morphologies and compositions as well as the valence state of compounds on the surface film were analyzed by SEM + EDS, and XPS.

The variation of wear scar width with load at 7min is shown in Figure 4.24. The variation of wear scar width with time at 5kg load is shown in Figure 4.25. The wear scar widths of FeS and FeS$_2$ coating were apparently less than that of original steel and sulfurized layer. With the increase of load, the wear scar widths of original steel and sulfurized layer increased, while the FeS (FeS$_2$) coating showed the contrary tendency. Figure 4.26 shows the variation of friction coefficient with time at 5kg load. The friction-reducing properties of the coatings were not as good as that of sulfurized layer due to their loose structure.

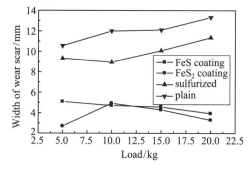

Fig.4.24   The variation of wear scar width with load at 7min

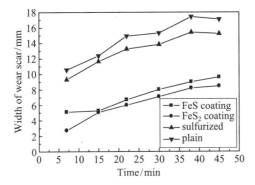

Fig.4.25   The variation of wear scar width with time at 5kg load

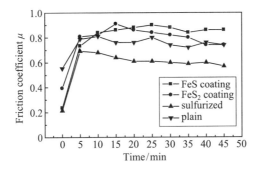

Fig.4.26   The variation of friction coefficient with time at 5 kg load

Figure 4.27 shows the worn morphologies of FeS (FeS₂) coating and sulfurized layer at 20kg load after sliding 7min. The compositions on the worn surfaces at different loads are shown in Table 4.6. Figure 4.28 shows the contents of sulfur and oxygen on the wear scar surface vs. width of wear scars. It can be found that, the higher the sulfur content on the worn surface, the smaller the width of wear scar, namely the better the wear-resistance. The content of oxygen on the worn surface had a certain effect on the wear-resistance. Less width of wear scar was formed when the content of oxygen was in the range of 10atom%—25atom%, the excessively higher or lower oxygen content was not good. The O/S concentration ratio on the worn surface vs. width of wear scar is shown in Figure 4.29. The width of wear scar was smaller at a higher concentration ratio. Based on the above testing results, it can be concluded that the wear-resistance was much better only when the sulfur content and the O/S concentration ratio were comparatively high, and the oxygen content was moderate. The contents of sulfur and oxygen were closely related to the experimental conditions. The sulfur content was high at a large load, while the oxygen content was not too high due to the competitive growth of sulfide and oxide during the frictional process. Table 4.7 shows the XPS analysis results. The surface films on the wear scars of FeS coating and original steel were all composed of iron sulfide or iron oxide.

(a) FeS coating   (b) FeS$_2$ coating

(c) sulfurized coating

Fig.4.27   Worn morphologies of the coatings at 20kg load after sliding 7min

Table 4.6   Compositions on the worn surface at different loads

Samples	Atom%	5kg	10kg	15kg	20kg
FeS coating	S	0.65	0.88	1.23	1.23
	O	18.13	23.20	13.57	18.16
FeS$_2$ coating	S	2.06	0.51	0.74	2.29
	O	11.63	15.62	20.25	18.72
Sulfurized layer	S	0.01	0.10	0.02	0.10
	O	4.05	27.92	20.21	28.37

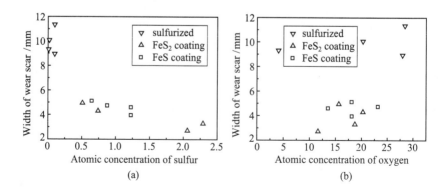

Fig.4.28   Contents of (a) sulfur and (b) oxygen on the worn scar surface vs. widths of wear scars

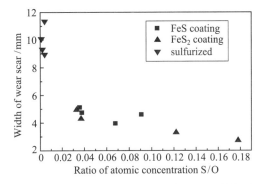

Fig.4.29   The O/S concentration ratios on the worn surface vs. width of wear scar

Table 4.7   The XPS analysis results

Load	Samples	Bonding energy			Compound
		S(2p)	O(1s)	Fe(2p)	
20kg	Original steel	—	530.1	709.2	FeO
	FeS coating	161.7	530.1	710.1	FeS, Fe$_2$O$_3$ or FeO
	FeS$_2$ coating	161.2, 162.5	530.2	710.1	FeS$_2$, FeS, Fe$_2$O$_3$ or FeO
5kg	Original steel	—	530.0	710.1	Fe$_2$O$_3$ or FeO
	FeS coating	161.5	530.1	710.8	FeS, FeO, Fe$_2$O$_3$
	FeS$_2$ coating	162.5, 161.0	530.1	710.9, 709.0	FeS, FeS$_2$, Fe$_2$O$_3$ or FeO

Figure 4.30 shows the worn morphologies of the steel rings at 20kg load after sliding 7min. Figure 4.31 shows the distribution of sulfur and oxygen elements on the

(a) Worn morphology of the steel ring rubbing with FeS coating

(b) Worn morphology of the steel ring rubbing with FeS$_2$ coating

(c) Worn morphology of the steel ring rubbing with original steel

Fig.4.30   Worn morphologies of the steel ring at 20kg load after sliding 7min

worn scar. At this moment, the surface of steel ring had been worn severely. There were high contents of sulfur and oxygen elements on the worn surface. Table 4.8 shows the content of sulfur and oxygen on the steel rings at 20kg and 5kg load. It can be found that the sulfur was transferred to the counterpart in the process of wear, and more sulfur was transferred at a high load. The content of oxygen on the steel ring rubbing with coating was higher than that rubbing with original steel, indicating that the content of oxide formed on the steel ring was promoted by sulfide. By the combined role of sulfide and oxide, the tribological properties of the coatings were improved.

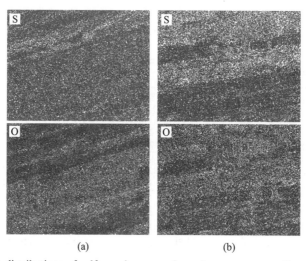

(a)   (b)

Fig.4.31   The distributions of sulfur and oxygen elements on worn scar of steel ring rubbing with (a) FeS coating (b) FeS$_2$ coating

Table 4.8   The contents of sulfur and oxygen on the steel ring at 20kg and 5kg load

Samples	Atom%	20kg	5kg
Steel ring rubbing with original steel	S	—	—
	O	8.53	6.13
Steel ring rubbing with FeS coating	S	4.86	0.73
	O	34.17	22.02
Steel ring rubbing with FeS$_2$ coating	S	8.30	7.82
	O	29.24	26.96

4.3   Plasma Sprayed Nano-FeS and FeS-SiC Composite Coating

Owing to the grain size and surface effects, nano-materials exhibit special performances of mechanics, electricity, thermology, corrosion- and wear-resistance, etc. The nano-material coatings show superior tribological properties to the traditional coatings. In recent years, there is an increasing interest in the studies of nano-coatings and nano-films. The researches showed that WC/Co, ZrO_2 and Al_2O_3/TiO_2 nano-coatings can significantly improve the wear-resistance of materials.

4.3.1   Plasma Sprayed Nano-FeS Coating

Table 4.9 shows the friction coefficient and wear scar width of nano-FeS coating and 52100 steel under dry friction condition. It can be found that both the friction coefficient and wear scar width of nano-FeS coating were lower than those of 52100 steel.

Table 4.9   The tribological properties of nano-FeS coating and 52100 steel under dry friction condition [4]

Load /N	Friction coefficient		Wear scar width /mm	
	FeS coating	GCr15 steel	FeS coating	GCr15 steel
75	0.3	0.50	1.6	2.10
125	0.4	0.58	2.4	2.80

The nano-FeS coating also possessed better tribological performance under oil lubrication condition as shown in Figure 4.32. The friction coefficient of FeS coating was nearly half that of 52100 steel, and the wear scar width was also obviously lower than that of 52100 steel.

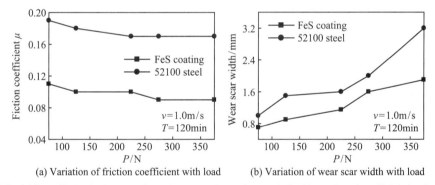

(a) Variation of friction coefficient with load   (b) Variation of wear scar width with load

Fig.4.32   Tribological properties of nano FeS coating and 52100 steel under oil lubrication condition [4]

4.3.2   Sprayed FeS-SiC Composite Coating

In order to further enhance the tribological performance of FeS coating, the FeS-SiC composite coating was sprayed using the mixed SiC and FeS particles [3]. The variations of wear loss and friction coefficient with time are shown in Figure 4.33. With the increase of SiC the friction coefficients all descended at different temperatures. Comparing with FeS coating, the friction coefficient of FeS-SiC composite coating decreased above 10%. The wear loss of the composite coating also exhibited the lowest value at 10%—20% SiC content below 350°C. In brief, the composite coating possessed the best friction-reducing behavior at SiC content of 10%—20%, and greatest wear-resistance below 350°C.

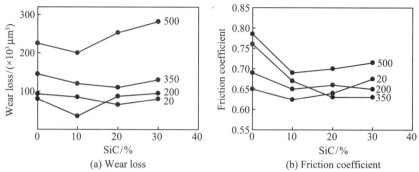

(a) Wear loss   (b) Friction coefficient

Fig.4.33   Variations of wear loss and friction coefficient of composite coatings with content of SiC [5]

4.4   Comparison of the Tribological Properties of Ion Sulfurized Layer and Plasma Sprayed FeS Coating

4.4.1   Experimental Method

The substrate was 1045 steel with hardness of HRC 55. The surface roughness was R_a= 0.04mm [6]. The purity of FeS particles for plasma spraying was 85%, their granularity was 30μm. For protecting the FeS particles from burning in the course of plasma spraying, they were covered by a layer of nickel-based alloy. The thickness of the spraying FeS coating was about 200μm. The ion sulfurized coating was prepared by low temperature ion sulfuration technique for 3h; its thickness was about 20μm. The friction and wear tests were carried out on a ball-on-disc testing machine of model QP-100. The upper sample was the 52100 steel ball with hardness of HV 770, and the lower samples were 1045 steel discs with coating. All tests were carried out under the lubrication condition with No.40 machine oil. The load of 70N and velocity of 1.60m/s were adopted for testing the variation of friction coefficient with time; the load of 82N and velocity of 1.49m/s for testing the variation of wear scar widths at time of 7.5min, 15min, 22.5min, 30min and 37.5min. For testing the variation of scuffing load with velocity, the variables were 1.12m/s, 1.49m/s, 1.87m/s, 2.25m/s and 2.70m/s.

4.4.2   Microstructure and Tribological Properties

Figure 4.34 shows the surface morphologies of ion sulfurized layer and plasma sprayed FeS coating. The former was relatively compact, while the latter was more porous and

(a) Ion sulfurized layer   (b) Sprayed FeS coating

Fig.4.34   The surface morphologies

incompact.

Figure 4.35 shows the X-ray diffraction patterns of ion sulfurized layer and plasma sprayed FeS coating. The phase structure of the former was relatively simple, whereas that of the latter was more complicated. The nickel oxide and two kinds of iron oxide were produced, showing that although the FeS particles were covered by nickel-base alloy, they were still oxidized and burned during spraying operation. The FeS phase appeared in both coatings, but the intensity of FeS peak of sprayed FeS coating was higher than that of ion sulfurized layer. It indicated that the contents of FeS in sprayed coating were much more than that in ion sulfurized layer.

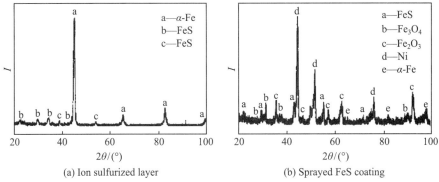

(a) Ion sulfurized layer   (b) Sprayed FeS coating

Fig.4.35   XRD patterns of two kinds of FeS coatings

Figure 4.36 (a) shows the variation of friction coefficient with time of ion sulfur-

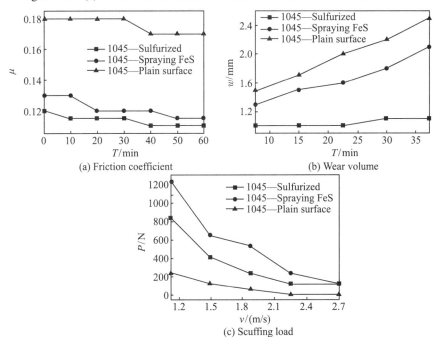

(a) Friction coefficient   (b) Wear volume

(c) Scuffing load

Fig.4.36   Tribological properties of the two kinds of FeS coatings

ized coating, sprayed FeS coating, and original 1045 steel surface. The friction co-efficient of all surfaces was stable and descendent slightly with time. The friction coefficient of ion sulfurized layer and sprayed FeS coating was obviously lower than that of original 1045 steel. Comparatively the friction coefficient of ion sulfurized layer was lower.

Figure 4.36 (b) is the variation of wear loss with time. The wear loss of two solid lubrication coatings were all less than that of original 1045 steel. Relatively the wear loss of ion sulfurized layer was lower.

Figure 4.36 (c) shows the variation of scuffing load with velocity. The two coatings possessed higher scuffing load at low and middle velocities. Along with the increase of velocity, scuffing load descended rapidly. The anti-scuffing property of the two solid lubricant coatings were all superior to that of original 1045 steel, whereas the sprayed FeS coating was much better than that of ion sulfurized layer.

4.5   Sol-gel FeS Coating

As a simple and practical method of preparing coatings, sol-gel technology has been applied widely. TiO_2 and Al_2O_3 wear-resistant films have been prepared by sol-gel process on the glass substrate. ZrO_2 wear-resistant coating doped with different rare earth, La_2O_3, $SrBi_2Ta_2O_9$, and oxide coatings have also been fabricated by this process.

Sol-gel process has many characteristics compared with other technologies.

(1) The process is simple and flexible; it doesn't need the vacuum condition or expensive vacuum equipment.

(2) The process is conducted under low temperature. It is especially important for preparing multicomponent system containing volatile components or easily inducing phase separation under high temperature.

(3) The films or coatings can be prepared on a variety of substrates with different shapes and materials, even the covering film can be fabricated on the particle surface of powder material.

(4) The uniform and multi-component oxide coatings can be prepared easily and the quantitative doping is also easily realized. In addition, the composition and microstructure can be controlled effectively.

4.5.1   Preparation

The substrate material was AISI 1045 steel with a hardness of HRC 55. The surface roughness was R_a =3.2μm. The purity of FeS powders was 85%, granularity was 2—3μm. The colloidal solution (solvent) was a patented inorganic silicon water-based paint (ZM-1). It has many advantages, such as high strength, low stress, natural drying, long life, and pollution-free. The filler materials (solute) were a mixture of FeS powders and zinc powders with diameter of 5μm. When preparing the sol, 500mL solvent and 100g solute were measured and mixed, 15min stirring was needed to form a homogeneous solution, in which FeS was used as solid lubricant and Zn was used to disperse evenly the FeS powder and enhance the wet ability of FeS powder to the substrate. To compare the difference of their tribological properties, four kinds of sol with different FeS contents were prepared. The FeS contents were 10g, 20g, 30g and 40g, respectively, the zinc contents were 90g, 80g, 70g and 60g. A special spraying gun was utilized to spray the sol homogeneously onto the surfaces of 1045 steel about 500μm thick. These coatings were dried in atmosphere for 2h.

4.5.2   Characterization

4.5.2.1   Morphologies

Figure 4.37 (a) shows the surface morphology of a sol-gel coating with 30% FeS particles [7]. The coating was relatively compact. The other three FeS coatings possessed very similar characteristics. Figure 4.37 (b)—(d) are the distribution maps of Zn, Fe and S elements under the same visual field, displaying that FeS grains were distributed evenly in the coatings. Figure 4.37 (e) is the EDS analysis result for Figure 4.37 (a),

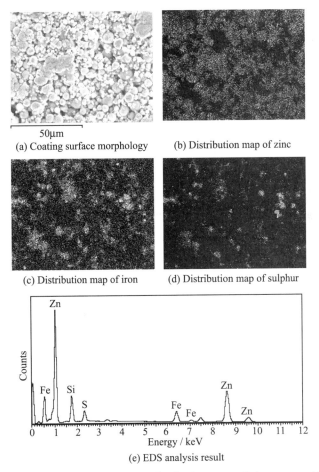

50μm
(a) Coating surface morphology   (b) Distribution map of zinc

(c) Distribution map of iron   (d) Distribution map of sulphur

(e) EDS analysis result

Fig.4.37   Morphologies of the FeS coating, distribution maps of elements, and EDS analysis result

showing that the intensity of Zn was much higher than that of S. The higher Si peak intensity came from the solvent.

4.5.2.2   Phase structure

Figure 4.38 shows the X-ray diffraction pattern of the coating with 30% FeS content, which is nearly similar to that of other three concentration coatings. Zn was in the form of simple substance and did not react with other elements in the sol. The intensity of Fe was very low, because the FeS coating was thick enough, covering the iron signal of 1045 steel substrate.

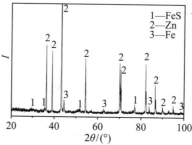

Fig.4.38   The phase structures of the coating

4.5.2.3   The bonding strength between coating and substrate

Figure 4.39 shows the bonding strength between coatings and substrates. The strength

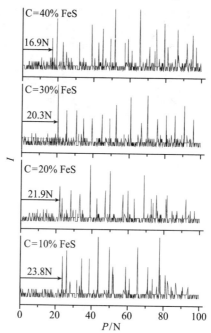

Fig.4.39   Bonding strength between the coatings and substrates

was generally lower due to the lower surface energy of FeS grains. The strength was decreasing when the FeS content increased.

4.5.3　Tribological Properties

Friction and wear tests were carried out on a ball-on-disc tester (QP-100 Model) under oil lubrication condition. The experimental parameters were the same as that of high velocity flame sprayed FeS coating. Figure 4.40 shows the variation of tribological properties of the FeS coatings and original 1045 steel under oil lubrication condition. The friction coefficients were all steady with increasing of time. However, the friction coefficient of the FeS coatings was obviously lower than that of original 1045 steel. The coating with 30% FeS content possessed the best friction-reducing performance.

The wear loss of the FeS coatings was less than that of original 1045 steel. The anti-scuffing property of four FeS coatings was superior to that of the original 1045 steel. Similarly, the coating with 30% FeS content showed the best wear-resistance and anti-scuffing load.

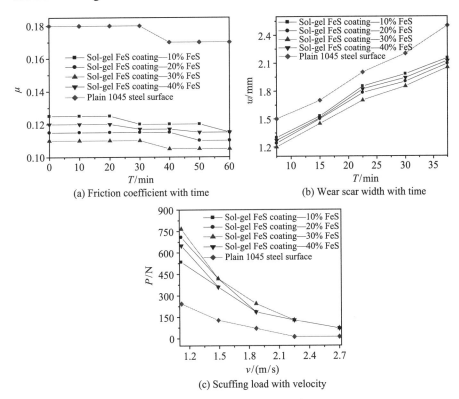

Fig.4.40　Variation of tribological properties of the FeS coating

Figure 4.41 shows the variation of friction coefficient with load, under fixed velocity 1.12m/s and 1.49m/s, respectively. The friction coefficient increased with velocity, indicating that the coating was damaged easily at the higher velocities. The friction coefficients also increased with load, whereas the FeS coatings possessed lower coefficients compared with 1045 steel.

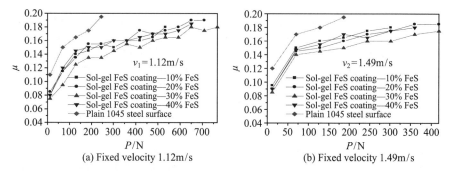

Fig.4.41   Variation of friction coefficient with load

Figure 4.42 shows the identification of elements on the spherical particles of the coa-ting with 30% FeS content. Figure 4.42 (a) is a selected area on a spherical particle, and Figure 4.42 (b) is the EDS analysis result for Figure 4.42 (a). The intensity of Zn was quite higher, while that of S, Fe and Si was nothing or extremely weak, indicating that these spherical particles were zinc simple substance.

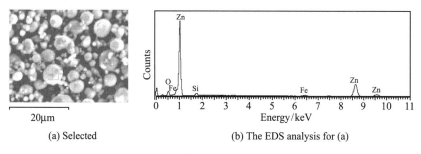

Fig.4.42   The identification of spherical particles on the coating surface

Figure 4.43 shows the worn morphology of the coating with 30% content FeS after sliding 15s. The entire coating was not damaged, but the wear had taken place within the FeS coating. Although zinc is also a solid lubricant, the lubrication effect was barely visible. Observing carefully, it can be found that the wear scars on the surface of zinc particles were rigid wear caused by abrasion. The zinc neither slipped along the close-packed plane like FeS and other laminal materials, nor generated plastic deformation and adhered on the counterpart surface like soft metals. It can be concluded that only FeS played the major friction-reducing role in the sol-gel coating. Figure 4.43 (b) is the worn morphology after 3min in the same experiment. The particles had been flattened, and the wear of FeS coating was greater. But on the whole, the coating kept integrated and did not flake off.

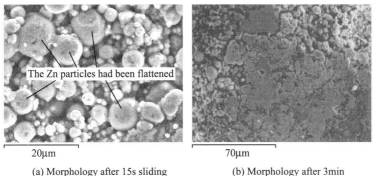

(a) Morphology after 15s sliding

(b) Morphology after 3min

Fig.4.43   Worn morphologies of the coating

Figure 4.44 (a) and (b) are the worn morphologies after 15min and 37.5min, re-spectively. The zinc particles no longer existed and had been polished. The longer the wear, the more smooth the surface. Figure 4.44 (c) and (d) are the EDS analysis results for Figure 4.44 (a) and (b), respectively. The intensity of elements was similar to that of unworn FeS coating, indicating that the FeS coating was surely not worn through and the wear only took place within the coating.

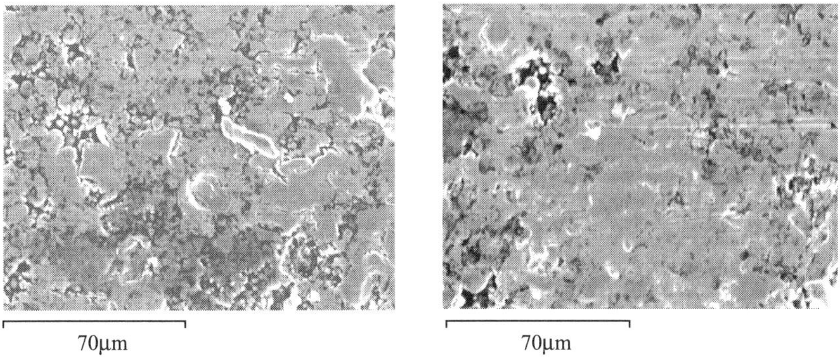

70μm

70μm

(a) Morphology after sliding 15min

(b) Morphology after sliding 37.5min

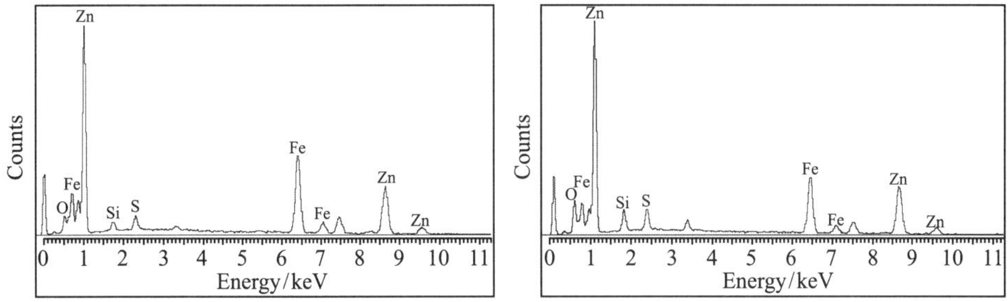

(c) EDS analysis for (a)

(d) EDS analysis for (b)

Fig.4.44   Morphologies and EDS analysis result of FeS coating

Figure 4.45 shows the surface morphology of the coating with 30% FeS content when scuffing happened. The major damages were fracture and flaking off. Figure

4.45 (b) and (c) are the EDS analysis results for the worn surface without spalling and worn surface after peeling. The elements distribution of unspalling worn surface was similar to the coating without wear, while that of exposed worn surface showed a characteristic of 1045 steel, demonstrating that the entire flaking off was the main damage form at the moment of scuffing.

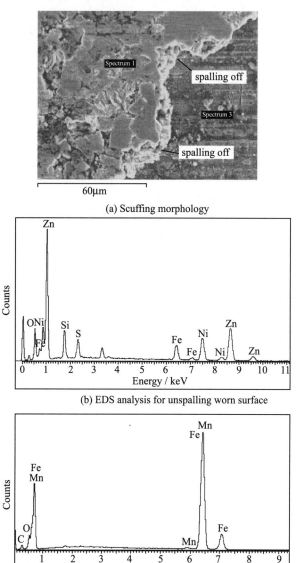

(a) Scuffing morphology

(b) EDS analysis for unspalling worn surface

(c) EDS analysis for exposed worn surface

Fig.4.45   Scuffing morphology and EDS analysis

Figure 4.46 shows the worn morphology and the EDS analysis of the steel ball.

There exists sulfur obviously on the worn surface, indicating that FeS had transferred to the counterpart, but Zn did not.

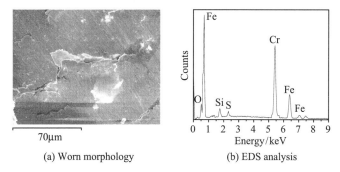

(a) Worn morphology	(b) EDS analysis

Fig.4.46   Worn morphology and EDS analysis of the steel ball

The tribological properties of four FeS coatings displayed the saddle-shape change with the increase of FeS content. When the FeS content increased from 10g to 30g, the tribological properties were improved. Whereas increasing to 40g, the effect went down. This phenomenon can be explained as follows: The solvent content of sol is constant (500mL), the total weight of solute, FeS and Zn powder, is also constant (100g), but the ratio between FeS and Zn are changeable. When the FeS contents are less than 20g, due to the lesser amount of FeS, the solid lubrication effect of the coatings is weak. When FeS content is up to 40g, due to the less content of Zn, the dispersibility and wettability of FeS powder are getting worse and the bonding strength between FeS coating and substrate descends rapidly. Therefore, the coating will flake off easily under the vertical pressure.

4.6   FeS Film Prepared by S-ion Implantation

4.6.1   Ion Implantation Technology

Since 1970s, ion implantation used for material surface modification has gradually been developed. It is a process to introduce the required ions into the surface of workpieces by applying super-high voltage. A large number of researches have confirmed that many properties of materials, such as friction-reduction, wear-resistance, oxidation-resistance, corrosion-resistance, and fatigue resistance, can be improved significantly by this technique.

4.6.1.1   Characteristics of ion implantation

(1) Ion implantation is not a heat balance process. The ions implanted possess high energy, which can reach 2—3 orders of magnitude higher than heat balance state. Therefore, in principle, each element in the periodic system of elements can be implanted into any matrix material.

(2) The kind, energy and content of implanted elements all can be selected. The surface alloy formed by implanting is not limited by the classical thermodynamic

parameters of diffusion and solubility, namely, new alloy phases can be obtained by this method.

(3) There is no obvious interface between the implanted layer and substrate material, and the bonding strength is very strong. Therefore, the adhesion breakdown or spalling of the layer will not occur.

(4) The concentration and depth of implanted ions can be controlled through adjusting the content, energy, and beam density.

(5) Ion implantation is carried out in vacuum at room temperature. No deformation and oxidization occurred on the workpiece. Therefore, the workpieces can keep the original accuracy and surface roughness; it is suitable for high-precision components as the final process.

(6) The compressive stress can be produced on the workpiece surface to reduce the initiation and propagation of cracks.

4.6.1.2 Strengthening mechanism of ion implantation

The hardness, wear-resistance, fatigue-resistance, corrosion-resistance and oxidation-resistance of implanted materials can be improved significantly; the mechanisms can be summarized as follows.

(1) Radiation damage strengthening.

The implanted ions with high energy will collide with substrate atoms so that a large amount of lattice damage can be induced. When the obtained energies of lattice atoms surpass the bonding energy of the lattice atoms, the displacement and vacancies, namely interstitial atoms will be generated. A strong radiation damage region is produced within the surface layer due to a series of collision cascade to form a great deal of dislocation defects.

(2) Solution strengthening.

As the ion implantation is a non-equilibrium process, supersaturated solid solution is easy to be formed. Therefore, the solution strengthening effect of materials containing excessive implanted ions is significant.

(3) Dispersion strengthening.

Due to the fierce atom collision and great target temperature-rise during the ion implantation process, precipitated phases will be formed to strengthen the matrix. For instance, when carbon, nitrogen, or boron atoms are implanted into steels, a series of iron compound precipitates, like carbon-iron, nitrogen-iron, or boron-iron compounds, can be formed.

(4) Grain refinement strengthening.

The fierce collisions of implantation will lead to the grain refinement of the substrate. As grain boundaries are the obstacles to dislocation movement, with the increase of grain boundaries, the movement of dislocation will be more difficult.

(5) Amorphous strengthening.

Amorphous metallic glass is free of dislocation and grain boundary, thus it possesses effective wear-resistance, oxidation-resistance and corrosion-resistance properties. Dozens of amorphous states can be obtained by ion implantation.

(6) Priority sputtering strengthening.

Different elements in alloy have different bonding energy. As the sputtering coefficient is inversely proportional to bonding energy, the alloy with weak bonding energy will be sputtered firstly. On the other hand, the alloy with high bonding energy possesses better strengthening characteristics. Therefore, the material surface can be strengthened by sputtering effect of implantation.

4.6.2   Tribological Properties of Sulfur-implanted Steel

Solid lubricant FeS was formed on a low-alloy steel surface by sulfur ion implantation [6]. The tribological properties had been investigated under boundary lubrication condition and compared with those of untreated steel. The pin-on-disc tests were carried out using a standard device at the condition of sliding speed 1.7m/s and load 15N.

The results showed that an obvious increase of average sliding length was obtained with the rise of S concentration. A high-implanted (the concentration of FeS = 30%) steel disc could run approximately 3 times longer compared with an untreated steel, as shown in Figure 4.47. There was also a significant effect even for the small concentration of 1%.

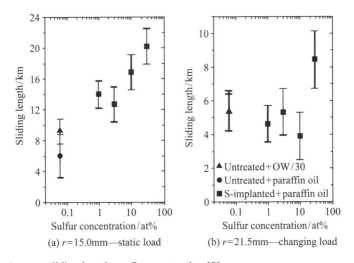

Fig.4.47   Average sliding length vs. S concentration [8]

References

1. Zhuang J. HVOF—The new member in the thermal spraying filed [J]. Locomotive & Rolling Stock Technology, 1997, (6): 28–31.
2. Zhang J, Jie X H. HVOF technology and its applications [J]. Heat Treatment of Metals Abroad, 1998, (2): 43.
3. Chen L M, Li Q. The present status and development of plasma spraying technology [J]. Heat Treatment Technology and Equipment, 2006, 27(1): 1–5.

4. Guan Y H, Xu Y, Zheng Z Y, et al. Tribological properties of nanostructured iron sulfide coating deposited by plasma spraying [J]. Tribology, 2006, 26(4): 320–324.
5. Guan Y H, Xu Y, Zheng Z Y, et al. Study on the microstructures of nano FeS-SiC composite coating by plasma spraying [J]. Transactions of Materials and Heat Treatment, 2005, 26(6): 105–108.
6. Wang H D, Zhuang D M, Wang K L, et al. Comparison of the tribological properties of an ion sulfurized coating and a plasma sprayed FeS coating [J]. Materials Science and Engineering: A, 2003, 357(1-2): 321–327.
7. Wang H D, Zhuang D M, Wang K L, et al. Study on tribological properties of iron sulfide coating prepared by sol-gel method [J]. Journal of Materials Science Letters, 2003, 22(20): 1603–1606.
8. Petersen J H, Reitz H, Benzon M E, et al. Tribological properties of sulfur-implanted steel [J]. Surface and Coatings Technology, 2004, 179: 165–175.

Chapter 5
Micron-nano MoS$_2$ Solid Lubrication Film

The researches on the MoS$_2$ film as the representative of solid lubrication has been paid great attention. Until now, a large number of methods have been developed to prepare MoS$_2$ films, such as RF sputtering, pulsed DC sputtering, magnetron sputtering, reactive sputtering, ion-beam assisted deposition, pulsed laser deposition, chemical vapor deposition, electrochemical deposition, and electro-deposition. In addition to the deposition films, MoS$_2$ bonding films are also applied extensively.

5.1 MoS$_2$ Film

5.1.1 MoS$_2$ Sputtering Film

In 1969, MoS$_2$ sputtering technology was developed by Spalvin. It is a process of applying high-energy plasma in vacuum under the action of electrical field to deposit the solid MoS$_2$ on the substrate surface. Before deposition, the ion bombardment is used to clean and activate the substrate surface so that the prepared coating is very dense with high purity and good bonding strength with the substrate. This kind of film doesn't need the adhesive; the coating is very thin (ranging from hundreds angstroms to one micron); the film thickness can be well controlled and the friction and wear properties are excellent. Therefore it can overcome the shortcomings of the bonding and electrochemical films, which can not ensure the precision of the precise equipment due to the bigger thickness and more debris produced. Meanwhile, when the bonding film is used in the environments like vacuum, high temperature, and radiation, etc., the adhesive will be volatilized or decomposed to gases, which can interfere the normal working of the precision instruments, and optical components, etc. In addition, the lubrication film will lose its effect due to the degeneration of the adhesive. However, the sputtering film can avoid the occurrence of above phenomena. The sputtering technology has opened a new way for the research and application of MoS$_2$ films and attracted the worldwide attentions.

From the 1970s, Lanzhou Institute of Chemical Physics had started the researches on the sputtering MoS$_2$ film. Until the 1980s, they adopted the improved equipment to apply the sputtering MoS$_2$ film to ball bearings used in vacuum. The MoS$_2$ film as the lubricant for the C36100 ball bearing (only the roller conveyer was coated) possessed low and stable friction moment, which started to increase until 2.5×10^6 cycles.

The sputtering MoS$_2$ film presents dark gray color; its thickness is ranging from 0.6μm to 0.9μm. As the difference of sputtering coefficient of Mo and S, the ratio of S/Mo is between 1.6 and 1.9. An interface region with elements staggered can be formed between the film and substrate and there is strong bonding strength between them. This film shows an amorphous structure with bigger layer space, containing a small amount of microcrystals.

5.1.1.1   Growth characteristics of the sputtering MoS₂ film

The lubrication behavior of the MoS_2 film is related to its structure [1]. It was found that when the plane (002) of the MoS_2 film parallel to the substrate surface, the friction coefficient is very low (<0.04), the wear-resistance is good, and the oxidation resistance is also very high; however, when the plane (002) is perpendicular to the substrate surface, the friction coefficient can achieve 0.4 and both the wear- and oxidation-resistance are very poor. Therefore, it is very important to prepare the MoS_2 film with a structure parallel to the substrate surface for obtaining good tribological properties.

The plane (002) of MoS_2 film prepared in most experimental conditions is perpendicular to the substrate surface. The parallel structure can be obtained only in strict conditions.

1. Working pressure

In the low pressure condition ($\leqslant 0.40Pa$), only the peak (002) appeared, while the peaks of (100) and (110) didn't, which demonstrated that the plane (002) was parallel to the substrate surface, as shown in Figure 5.1. When the working pressure was greater than 0.40Pa (like 0.6Pa), the peaks of (002), (100) and (110) all appeared; and with the increase of working pressure, the intensity of (100) and (110) peaks was increased, the high intensity of (100) and (110) peaks means that the structure of MoS_2 film became less planes of (002) parallel to the surface.

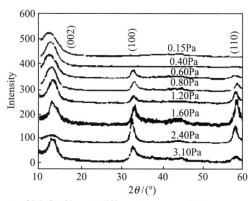

Fig.5.1   XRD patterns of MoS₂ films at different pressures [1]

2. Deposition time

At the same pressure of 0.4Pa, regardless of the deposition time, only the peak (002) appeared, indicating that the MoS_2 film grew parallelly to the substrate surface, as shown in Figure 5.2 (a). When the pressure was up to 0.88Pa and 1.60Pa, in the early stage of deposition, the appeared peak of the two films was mainly (002); with the increase of deposition time, the (100) and (110) orientations of the films began to be dominative, indicating that the structure of MoS_2 film changed from parallel to perpendicular to the substrate surface.

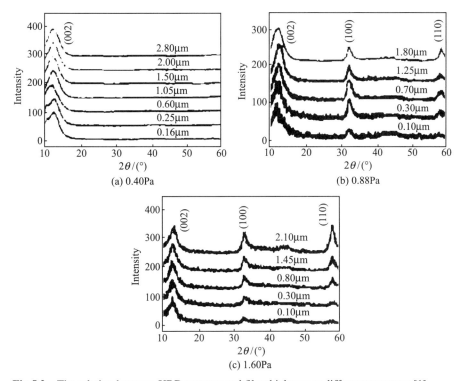

Fig.5.2   The relation between XRD patterns and film thickness at different pressures [1]

The growth of MoS₂ film not only depended on the working pressure and deposition time, it was also related to the deposition rate and deposition particle energy. When the deposition rate was lower; deposition particle energy was higher, the MoS₂ film was easy to be formed with a smooth, dense and (002) surface-oriented layer structure; conversely, the MoS₂ film would be formed with a rough, loose, and joint growth of (002) and (100) or (110) structure.

5.1.1.2   Tribological properties of the sputtering MoS₂ film

The friction coefficient of AISI 52100 steel substrate was high reaching about 0.9 as shown in Figure 5.3 (a), while the friction coefficient of the MoS₂ film was decreased to about 0.2 (Figure 5.3 (b)), indicating that the sputtering MoS₂ film possessed excellent friction-reducing behavior [2].

5.1.2   MoS₂ Film Prepared by Two-step Method

The sputtering MoS₂ film can effectively reduce the friction, wear and scuffing of friction-pairs, but because of some weakness: like uneven thickness, rough surface, low deposition rate and inaccurate atomic ratio of S and Mo, the sputtering method is unsuitable to the precision friction-pairs. Some researchers employed a two-step method, namely multi-arc ion plating Mo and low temperature ion sulfuration, to prepare the solid lubrication MoS₂ films, and then their microstructures and tribological properties were studied.

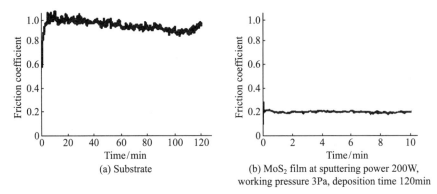

Fig.5.3 Curves of the friction coefficients of 52100 steel and the MoS₂ film [2]

5.1.2.1 Multi-arc ion plating Mo + sulfurizing combined treatment

1. Multi-arc ion plating technology [3]

Ion plating technique was developed in the early 1960s on the basis of vacuum evaporating and vacuum sputtering. Multi-arc ion plating is an improved method, which was firstly developed by the former Soviet Union. In the early 1980s, this technique was first pragmatized by the Multi-Arc Company of United States.

The structure of evaporation source of multi-arc ion plating is shown in Figure 5.4. It is composed of water-cooled cathode, magnetic coil, and ignition arc electrode, etc. The cathode is made from the coating material. Under the vacuum condition of 10—10^{-1}Pa, the power supply is given, the ignition electrode instantly contacts with cathode. As the conductive area decreases rapidly at the moment of leaving of electrode, the resistance increases; the temperature of local region quickly increases, which results in the melting of the cathode material to form liquid bridge conduction. Finally, the metal is evaporated explosively; the local region with high temperature is formed on the cathode surface; the plasma is generated to ignite the arc, and the power supply with low pressure and high current maintains the continuous arc discharge. Many bright and mobile spots are formed on the surface of cathode, namely the so-called arc spots of cathode. The spots with high current density and high-speed movement exist in the extremely small space; their sizes are ranging between 1μm and 100μm. Their current density can reach 105—107A/cm². The existence time of each spot is very short, after their explosive ionization to emit ions and electrons to evaporate the cathode material, the metal ions form a space charge at the vicinity of the cathode surface, the condition for generating arc spots is established again and new arc spots will be reproduced. A lot of arc spots continue to be generated, which may maintain the stability of the total arc current. The cathode material with the ionization rate of 60%—90% for each arc spot is evaporated and deposited on the substrate surface to form coating. The movement direction and speed of arc spots are controlled by the magnetic field. Appropriate magnetic field strength can make arc spots small, dispersive and etching the cathode surface uniformly.

The basic principle of arc ion plating is that the metal evaporation source (target source) is used as cathode; the target is evaporated and ionized through the arc discharge between the cathode and anode to form space plasma, by which the coating is deposited on the workpiece.

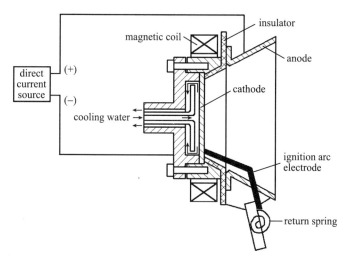

Fig.5.4   The schematic diagram of the arc ion plating with cathode mandatory cooling [3]

Arc ion plating possesses the following characteristics.

(1) As the evaporation source of cathode arc does not have a molten pool, it can be placed arbitrarily in a proper position in the plating chamber. A number of arc evaporation sources can be also employed, so that it can improve the deposition rate, make the film thickness homogeneous and simplify the rotating mechanism of substrate.

(2) The ionization rate is high and can reach more than 80%; therefore, the plating rate is very high, and it is beneficial to improve the bonding strength between the coating and the substrate as well as the coating performance.

(3) The arc has many functions. It is the evaporation and ionization source as well as the ion source of ion sputtering cleaning.

(4) The deposition speed is fast and the ion detour ability is good.

(5) The energy of incident particles is high; the strength and wear-resistance are good. The atom diffusion exists at the interface of the coating and workpiece, so that the adhesion of the coating is high.

2. Preparation [4]

The substrate materials used for experiments were monocrystalline silicon and 1045 steel. The former was used for micro-analysis, while the 1045 steel was used for friction and wear tests. The surface roughness of the crystalline silicon was $R=$ 0.2nm. The 1045 steel was heat-treated to the hardness of HRC 55, its surface roughness was $R_a=0.04\mu m$.

The metal Mo film was prepared in the multi-arc ion plating equipment of model MIP-6-800. Figure 5.5 shows the schematic diagram of MIP-6-800. Three molybdenum targets of $\phi150mm\times30mm$ were placed symmetrically in the vacuum chamber. The specimens were hung on the rotary trestle with uniform rotation near the Mo targets. When the background vacuum was up to the scheduled value, the inert gas Ar was induced and the negative bias was set up. Then the arc was generated to create the ion beam flow of molybdenum. Under the negative bias, the ion flow was accelerated to bombard the surface of substrate to form a Mo film. The deposition time was 2h.

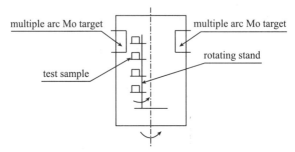

Fig.5.5   Schematic diagram of the multi-arc ion plating equipment

The prepared Mo film was then treated by low temperature ion sulfuration in the equipment of model DW-1. The workpieces (1045 steel coated with Mo film) were put on the cathode tray and the furnace chamber wall was linked to the anode. When the vacuum was up to a certain value, the ammonia gas was filled into the chamber. The direct current was applied between the cathode and anode to ionize the ammonia gas, the ammonia ions were accelerated to bombard the cathode (workpieces), elevating their temperature. Till the temperature was up to a scheduled value of 190°C, bombardment stopped. At this temperature the solid sulfur in the chamber became gasified, and the sulfur atoms permeated into the Mo film through the grain boundaries and crystal defects, forming the MoS₂ film.

3. Characterizations [4]

Figure 5.6 shows the surface morphology of synthetic MoS₂ film by SEM. The surface was loose and the grains size was large and inhomogeneous.

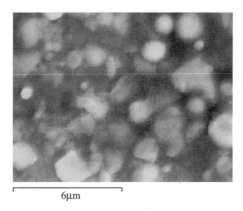

6μm

Fig.5.6   Surface morphology of the MoS₂ film by SEM

Figure 5.7 shows the surface morphologies of single Mo and MoS₂ film by AFM. In Figure 5.7 (a), the compact and ordered columnar grains were present on the surface of the Mo film, the grains size was over 100nm. Figure 5.7 (b) displays the obvious

shar- pisland shape grains. The grains size was under 100nm. This morphology characteristic is formed due to the bombardment of ammonia ions in the earlier stage of
sulfuration.

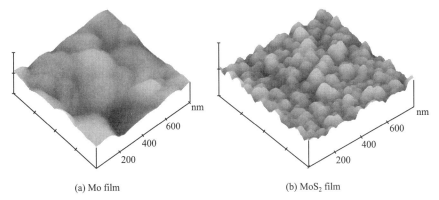

(a) Mo film   (b) MoS₂ film

Fig.5.7   Surface morphologies of the films observed by AFM

Figure 5.8 (a) is the cross-section morphology of the synthetic MoS₂ film by SEM,
showing the thickness of about 2μm. The bonding between the film and substrate was
compact and there was no transition layer at the interface. Figure 5.8 (b) is the enlarged
image of Figure 5.8 (a), in which the line distributions of oxygen, iron, molybdenum
and sulfur elements were labeled. There were a few Fe element in the film and Mo element in the substrate, which indicated that near the interface a little element transfer
happened. This phenomenon is because of that during ion plating Mo, the molybdenum ion beam flow continuously bombarded the 1045 steel surface, the Fe ions were
sputtered out and then co-deposited on the substrate surface with the molybdenum
ions. The atom diameter of Mo is 0.201nm, is close to that of Fe (0.172nm). Therefore, the Mo atoms could easily permeate into the Fe substrate. Figure 5.8 (b) shows
that the content of molybdenum was the highest at the outer layer of the film. The
content of sulfur was high in the surface, and decreased gradually from surface to inside. As part of sulfur could diffuse into the substrate, it might react with iron to form
FeS.

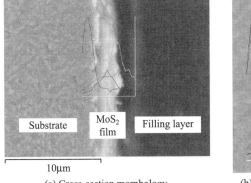 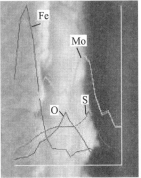

(a) Cross-section morphology   (b) Composition line scanning

Fig.5.8   Cross-section SEM image and composition line scanning of the synthetic MoS₂ film

Figure 5.9 shows the X-ray diffraction patterns of Mo and synthetic MoS₂ film. The Mo film had an obvious polycrystalline structure. As shown in Figure 5.9 (b), besides MoS₂, the FeS was also present showing that the sulfur ions permeated not only into the Mo film, but also into the 1045 steel substrate and reacted with Mo and Fe atoms to form MoS₂ and FeS. The highest single Mo peak indicated that although part of Mo atoms in the Mo film had been transferred to MoS₂, majority of them was still remained as single substance. It could be considered that the synthetic MoS₂ film prepared by two-step method was a metal-base composite film of Mo containing MoS₂.

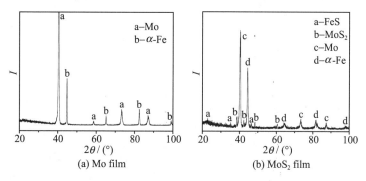

Fig.5.9   XRD patterns of the films

Figure 5.10 shows the chemical valence states of the molybdenum and sulfur elements on the MoS₂ film analyzed by XPS. It could be concluded that there were two phases present in the MoS₂ film. One was dominant single substance molybdenum, another was solid lubricant MoS₂.

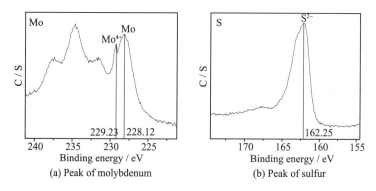

Fig.5.10   Analysis of the MoS₂ film by XPS

Figure 5.11 shows the AES composition depth profiles for the Mo and MoS₂ films. It could be found that molybdenum was evenly distributed along the depth from subsurface. The distribution trend of elements showed that the solid lubricant MoS₂ existed mainly in the near surface region of the film. In the sub-surface the single substance molybdenum was dominant and MoS₂ was subsidiary.

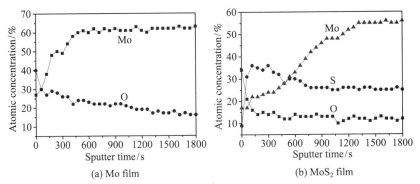

Fig.5.11   AES composition depth profiles (Ar^+ sputtering velocity: 30nm/min)

Figure 5.12 shows the bonding strength between the Mo, MoS₂ films and substrate, measured by a scratching tester. The peak of supersonic wave occurred suddenly meant that the bonding was destroyed. The bonding strength of them was sufficiently high up to 68.6N and 59.7N. The MoS₂ film showed lower bonding strength probably due to the formation of MoS₂, which disturbed the continuity of interaction between Fe and Mo atoms.

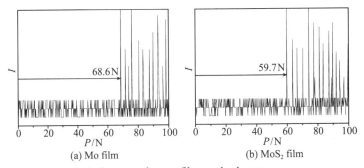

Fig.5.12   Bonding strengths between the two films and substrate

Figure 5.13 shows the nano-mechanical performance of the Mo layer and synthetic MoS₂ film, Figure 5.13 (a) is the nanohardness, and Figure 5.13 (b) is the nano-elastic

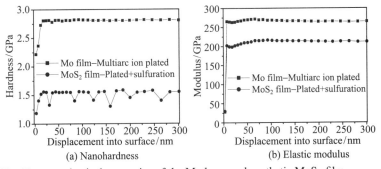

Fig.5.13   Nano-mechanical properties of the Mo layer and synthetic MoS₂ film

modulus. It can be found that the nano-hardness and nano-elastic modulus of the MoS$_2$ film are all decreased compared with those of Mo layer. The lower hardness and elastic modulus are beneficial to the friction-reduction.

4. Tribological properties [5−7]

The friction and wear tests were carried out on a ball-on-disc testing machine of model QP-100. The upper samples were 52100 steel balls with diameter 12.7mm and hardness HV 770, the lower samples were the 1045 steel discs with the synthetic MoS$_2$ film and dimension of ϕ60mm×5mm. As comparison, the 1045 steel and the FeS film with 6μm thickness (sulfurized 1045 steel) were also examined under the same experiment condition. The samples were all lubricated with engine oil, its kinematic viscosity was 37—43mm^2/s at 50°C and the velocity of oil dripping was 2mL/min. When testing the variation of friction coefficient with time, a load of 50N and velocity of 1.60m/s were fixed and the time changed from 0 to 60min. When testing the variation of worn scar widths with time, a load of 82N and velocity of 1.49m/s were fixed and the times were 7.5min, 15min, 22.5min, 30min and 37.5min. When testing the variation of worn scar widths with load, the time of 7.5min and velocity of 1.49m/s were fixed, and the loads were 24N, 82N, 140N and 198N. When testing the variation of scuffing load with velocity, the following velocity variables were employed 1.12m/s, 1.49m/s, 1.87m/s, 2.25m/s and 2.70m/s. At each velocity, the load was added stepwisely. The increment of load was 58N and the duration time was 2min.

Figure 5.14 shows the tribological properties of the synthetic MoS$_2$ film. It can be

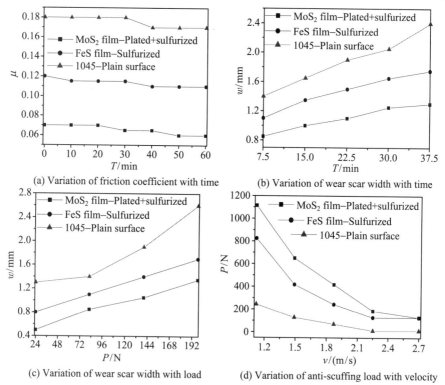

(a) Variation of friction coefficient with time

(b) Variation of wear scar width with time

(c) Variation of wear scar width with load

(d) Variation of anti-scuffing load with velocity

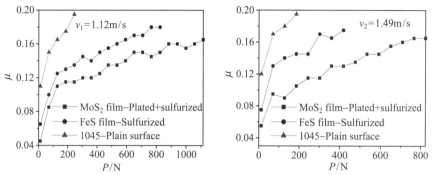

(e) Variation of friction coefficient with velocity (v_1) (f) Variation of friction coefficient with velocity (v_2)

Fig.5.14　Tribological properties of the synthetic MoS$_2$ film

seen that the friction coefficient of the synthetic MoS$_2$ film was lower about one time that of the FeS film, showing the excellent friction-reducing property.

The wear-resistance of MoS$_2$ film was also the best one among all samples. The scuffing load of the synthetic MoS$_2$ film was three to four times that of the FeS film and original 1045 steel at the velocity 1.12m/s. But along with the increase of velocity, the scuffing load descended rapidly.

Figure 5.15 shows the scuffing morphology of synthetic MoS$_2$ film and the distribution maps of Fe, Mo, S elements in the same field. The failure mode of MoS$_2$ film

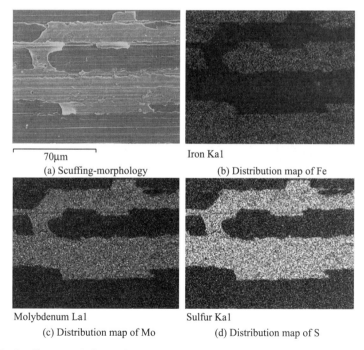

Fig.5.15　Scuffing-morphology of the synthetic MoS$_2$ film and the distribution maps of Fe, Mo, S elements in the same field

was ploughing and flaking off induced by the abrasive and fatigue wear.
Figure 5.16 shows the flaking off of MoS$_2$ film during scuffing.

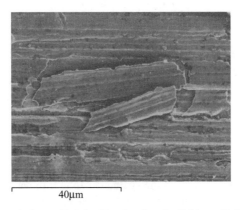

40μm

Fig.5.16   Scuffing morphology of MoS$_2$ film showing the flaking off of film

Figure 5.17 (a) and (b) show the worn morphologies of the synthetic MoS$_2$ film
before and after scuffing. Before the film was scuffed, the main wear mechanism was
abrasive wear, while the severe deformation and spalling by strain fatigue took place
after the film was scuffed. Figure 5.17 (c) and (d) are the EDX analysis for Figure
5.17 (a) and (b).

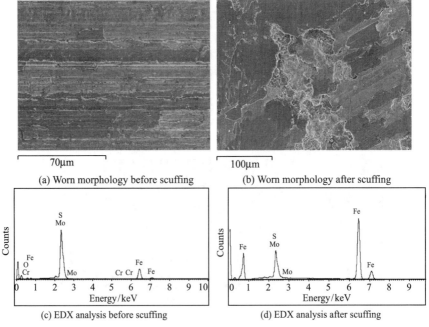

Fig.5.17   Worn morphologies and EDX analysis of the synthetic MoS$_2$ film before and after
scuffing

Figure 5.18 (a) and (b) are the AES composition depth profiles of the synthetic MoS$_2$ film before and after scuffing. Figure 5.18 (a) indicated that although the MoS$_2$ film was subject to a certain wear, it still played the lubrication role. In Figure 5.18 (b), the content of iron was obvious higher than that of molybdenum and sulfur, demonstrating that the MoS$_2$ film had been worn away.

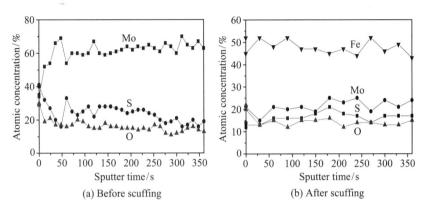

(a) Before scuffing   (b) After scuffing

Fig.5.18 AES composition depth profiles of the synthetic MoS$_2$ film

Figure 5.19 shows the XPS analysis results of the chemical valence states of iron, molybdenum and sulfur on the scuffed surface. It can be confirmed that the first strong peak was Fe; the second one was Fe^{2+} (FeO). The existence of Fe was caused by the spalling of the MoS$_2$ film exposing the substrate; the existence of FeO was due to that the steel ball directly contacted with the 1045 steel substrate and the iron reacted with oxygen to generate FeO under the friction heat. Figure 5.19 (b) is the XPS analysis result of molybdenum. The existence of the first two strong peaks indicated that the film had not flaked off entirely; the appearance of MoO$_3$ might be the product generated by the interaction of molybdenum and oxygen promoted by the friction heat. Figure 5.19 (c) is the XPS analysis result of sulfur. It was found that a strong S^{2-} peak occurred at the bonding energy 160.60eV, it could be concluded that the corresponding compound of the S^{2-} was MoS$_2$.

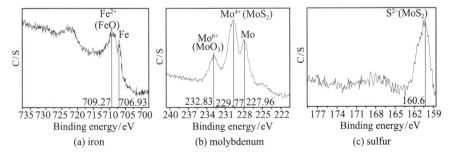

(a) iron   (b) molybdenum   (c) sulfur

Fig.5.19 XPS analysis results of the chemical valence states of iron, molybdenum and sulphur on the scuffed surface

Figure 5.20 (a) shows the worn morphology of the steel ball at 82N after sliding 7.5min. The surface was smooth with smaller wear scar. Figure 5.20 (b) is the EDS analysis results for Figure 5.20 (a). The obvious Mo and S peaks indicated that they had transferred from the disc to ball during the friction process. The transfer of the solid lubricant between the friction surfaces effectively improved the lubrication condition and enhanced the tribological properties of the friction-pair. Figure 5.20 (c) is the worn morphology of the steel ball at 1.87m/s sliding velocity when it was scuffed. It was found that severe plastic rheologic "adhesive wave" appeared on the scuffed surface; it was a typical morphology of adhesive wear. Figure 5.20 (d) was the EDS analysis result for Figure 5.20 (c). The Mo and S elements had almost disappeared.

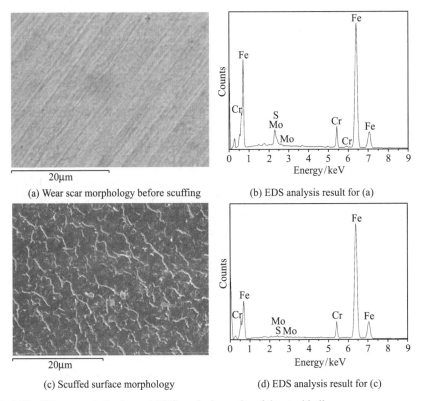

(a) Wear scar morphology before scuffing   (b) EDS analysis result for (a)

(c) Scuffed surface morphology   (d) EDS analysis result for (c)

Fig.5.20   Worn morphologies and EDS analysis results of the steel ball

5.1.2.2   Magnetron sputtering Mo + sulfurizing combined treatment

1. Preparation

The substrate material was 1045 steels heat treated to hardness of HRC 55. The average surface roughness was R_a=0.8μm.

The Mo layer was prepared in a magnetron sputtering equipment of model GDM-450BN. Figure 5.21 shows the schematic illustration of the magnetic sputtering system. The Mo target had a purity of 99.99%, with the dimensions of 124mm×254mm×

2mm. The workpieces were placed on the trestle. The temperature of the substrate during deposition was 100°C. The base pressure prior to deposition was 3×10^{-3}Pa, and a high-purity Ar gas (99.999%) was used as the sputtering gas. The current applied to the Mo target was a constant value 2A, the deposition rate of Mo layer measured was about 100nm/min. The thickness of Mo layer was 2μm; the deposition time was 20min.

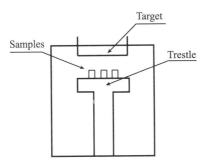

Fig.5.21 Schematic illustration of the magnetic sputtering system

The Mo layer was then treated by low temperature ion sulfuration for 2h.

2. Characterizations

Figure 5.22 (a) shows the surface morphology of the synthetic MoS₂ film. On the surface many spherical particles were present, whose size ranged between 50nm and 100nm. Figure 5.22 (b) shows the cross-section morphology and the distribution of Fe, Mo and S elements.

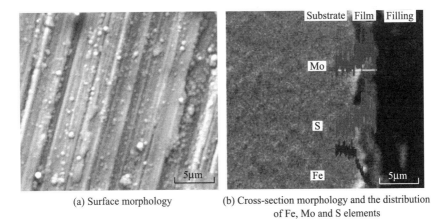

(a) Surface morphology   (b) Cross-section morphology and the distribution
                                     of Fe, Mo and S elements

Fig.5.22 Surface, cross-section morphologies of MoS₂ film and the distribution of Fe, Mo and S elements

Figure 5.23 is the XRD pattern of the synthetic MoS₂ film. The diffraction peaks were mainly Mo and MoS₂.

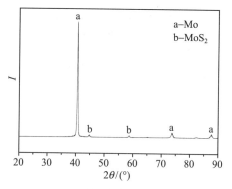

Fig.5.23 XRD pattern of the synthetic MoS$_2$ film

Figure 5.24 shows the load-displacement curves of the Mo layer and the synthetic MoS$_2$ film. The nano-hardness and elastic modulus of all samples are listed in Table 5.1.

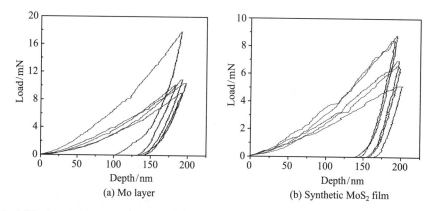

Fig.5.24 Load-displacement curves of the Mo layer and the synthetic MoS$_2$ film

Table 5.1 Nano-hardness and elastic modulus of all samples

Sample	1045 steel	FeS film	Mo layer	MoS$_2$ film
Nanohardness/GPa	10.51	2.84	13.26	7.44
Elastic modulus/GPa	247.76	52.97	334.68	244.55

3. Tribological properties

Friction and wear tests were conducted on a ball-on-disk wear tester of model T-11. The upper samples were 52100 steel balls with diameter of 6.35mm and hardness of HV 770. The lower samples were1045 steel disks with dimensions of ϕ25mm$\times$ 6mm, 1045 steel with MoS$_2$ and FeS films. Experimental parameters under dry condition were load 5N, sliding speed 0.2m/s, and experiment time 60min. Experimental parameters under oil lubrication condition were: load 20N, sliding speed 0.2m/s, and experiment time 2h.

Figure 5.25 shows the curves of tribological properties under dry condition. The variation of friction coefficient with time is shown in Figure 5.25 (a). The friction coefficient of the synthetic MoS$_2$ film was always lower than that of the FeS film and 1045 steel. Figure 5.25 (b) shows the variation of wear scar depth with time. The wear scar depth of the synthetic MoS$_2$ film was always lower than that of the FeS film and 1045 steel.

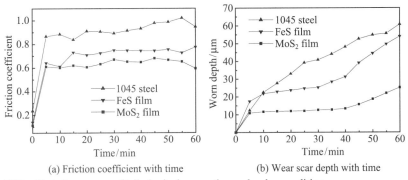

(a) Friction coefficient with time   (b) Wear scar depth with time

Fig.5.25 Variation curves of tribological properties under dry condition

Figure 5.26 shows the worn morphologies and compositions of the 1045 steel, FeS

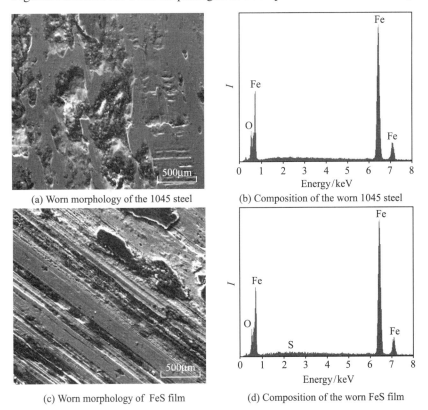

(a) Worn morphology of the 1045 steel   (b) Composition of the worn 1045 steel

(c) Worn morphology of FeS film   (d) Composition of the worn FeS film

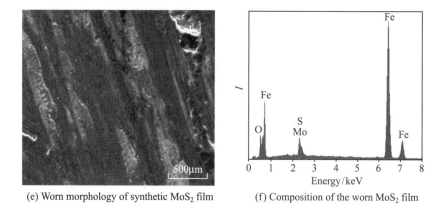

(e) Worn morphology of synthetic MoS₂ film　　　(f) Composition of the worn MoS₂ film

Fig.5.26　Worn morphologies and compositions of three samples under dry condition

film and MoS₂ film. Wide and deep wear scars and severe plastic deformation were generated on the 1045 steel surface showing the abrasive and adhesive wear mechanism. The wear surface of the synthetic MoS₂ film was comparatively smooth, and no adhesion happened. Meanwhile, it was found that no sulfur existed on the worn FeS film, it had entirely been destroyed; but a certain content of sulfur and molybdenum still existed on the worn MoS₂ film.

Figure 5.27 shows the variation curves of tribological properties under oil lubrication condition. The variation of friction coefficient with time is shown in Figure 5.27 (a). During the whole test, the friction coefficient of the synthetic MoS₂ film was always lower than that of the FeS film and 1045 steel. Figure 5.27 (b) shows the variation of wear scar depth with time. The wear scar depth of the synthetic MoS₂ film was very stable and always lower than that of FeS film and 1045 steel.

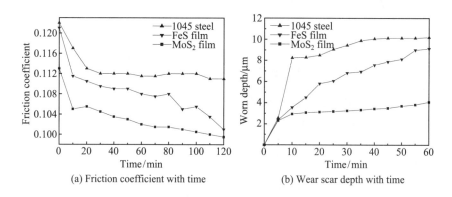

(a) Friction coefficient with time　　　(b) Wear scar depth with time

Fig.5.27　Variation curves of tribological properties under oil lubrication condition

Figure 5.28 shows the worn morphologies and compositions of the 1045 steel, FeS film and MoS₂ film under oil lubrication.

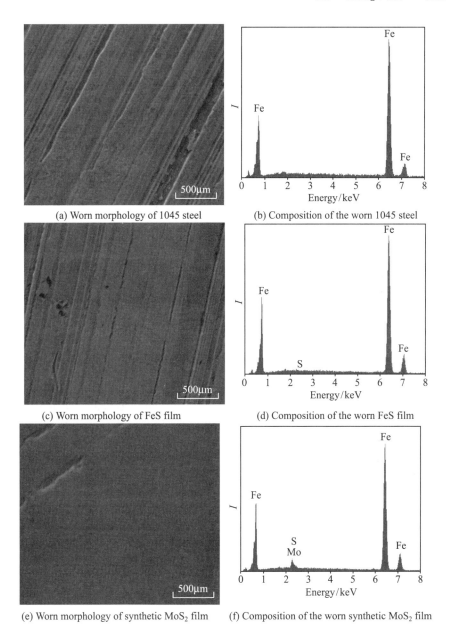

(a) Worn morphology of 1045 steel

(b) Composition of the worn 1045 steel

(c) Worn morphology of FeS film

(d) Composition of the worn FeS film

(e) Worn morphology of synthetic MoS₂ film

(f) Composition of the worn synthetic MoS₂ film

Fig.5.28   Worn morphologies and compositions of three samples under oil lubrication condition

5.1.2.3   Lubrication mechanism of synthetic MoS₂ film

MoS_2 possesses a close-packed hexagonal structure; its basic crystal lattice is the close-packed thin layer groups formed by three planes of S, Mo, S. The thickness of a unit group is 0.625nm, the slippage happens easily along the close-packed planes. In addition, the MoS$_2$ film is porous; it can store easily the lubricating oil to form a continuous oil film on the friction surface to prevent the direct contact between metals.

Due to the big differences of atom size, electrochemical factor and lattice structure between Mo and S, the sulfur atoms are hardly dissolved in molybdenum to form a homogeneous solid solution. The sulfur can permeate into molybdenum only through the grain boundaries and crystalline defects of the latter like the formation of FeS in sulfurizing process. Therefore, the synthetic MoS$_2$ film can be formed only at the outer sub-surface of the Mo layer. According to the tribological theory, an ideal friction surface should be soft on the surface, possessing excellent lubricating property, and hard in the sub-surface, giving a sufficient support to the lubrication film. The synthetic MoS$_2$ film is just such an ideal friction surface.

During the wear process, under the high load condition, the temperature of the contact area is very high after a long-time friction. Thus, MoS$_2$ can be oxidized in air and release the atomic sulfur, which can react with Mo and Fe to produce MoS$_2$ and FeS again. Therefore, the co-existence of molybdenum disulfide, iron sulfide, molybdenum oxide and iron oxide is very beneficial to improve the tribological performance.

5.1.3 Thermal Spraying MoS$_2$ Film

Figure 5.29 shows the SEM morphology of the nano-micron MoS$_2$ coating prepared by plasma spraying. The coating surface presented typical lamellar structure; each big flake was composed of small ones; the size of these small flakes ranged from 100nm to 400nm. It can be seen from the cross-section morphology (Figure 5.29 (c)) that the coating combined well with the substrate, and its thickness was relatively homogeneous.

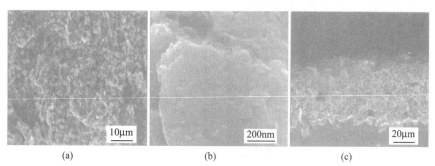

| (a) | (b) | (c) |

Fig.5.29 SEM morphology of the nano-micron MoS$_2$ coating [8]

Figure 5.30 shows the variations of friction coefficient and width of wear scar with load of the MoS$_2$ coating and 52100 steel under oil lubrication condition. The friction coefficient of each sample decreased with the increase of load, and it gradually approached to a stable value. The friction coefficient of MoS$_2$ coating was lower about one time that of the 52100 steel. The wear scar width of MoS$_2$ coating was obviously lower than that of the 52100 steel as well; especially in the comparatively higher load condition.

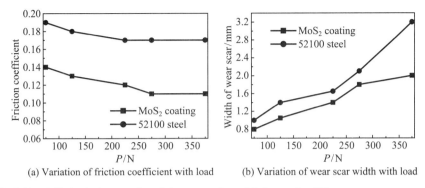

Fig.5.30 Tribological properties of the nano-micron MoS$_2$ coating [8]

5.1.4 Bonded MoS$_2$ Film

Bonded solid lubrication coatings, which are also known as dry film, possess many characteristics, such as simple preparation, low cost, and good friction-reducing performance. They have widely been applied to many fields, like civilian machinery and space components, for instance, for the fitting surface of blade rabbet of engine compressor and bolts.

The lubrication role of dry film bases on the transfer lubrication. When one surface of a friction-pair is sprayed with dry film, during operation, most part of it can be transferred to the counterpart surface making the friction occurred between the dry film and transferred film. At this condition, the lubrication state is the best; the friction coefficient is the lowest. However, with the extension of running, the dry film will be worn constantly; the wear debris particles become dry and hardened by the frictional heat, the dry film and transferred film are scratched, the wear speed is accelerated, which will finally lead to the failure of solid lubrication.

There are abundant dry film species, the industrial countries and the air companies all have their products. Since the 1970s, Lanzhou Institute of Chemical Physics has carried out the systematic researches on the solid lubrication materials. For more than 30 years, a variety of dry films such as silicate series, polyimide series, epikote, polyphenylene sulfide, and phosphate, etc. were developed. Some of them have successfully been applied to the lubrication of many components such as moving parts of manmade satellite, delivery pump bearings of missile fuel, torsion springs, hinges, and gears. SS-2 and IF-1 inorganic and organic dry films have been used to the aero space vehicles. In addition, they have also used polyphenylene sulfide to develop PPS-1 dry film, which had longer life than other organic films at room temperature; however, the appearance of the dry film was rough with pinholes; the wear-resistance was low when the film thickness was less than 70μm, and the film was easy to flake off. Therefore, on the basis of PPS-1 dry film, through improving the batch formula, they prepared PPS-2 dry film, which solved above problems, and extended the film life.

According to the studies of relevant researchers, solidification temperature and time had big impact on the wear-resistance of PPS-2 dry film [9]. Tables 5.2 and 5.3 show the effect of solidification temperature and time on the wear-resistance of PPS-2 dry film on the substrate of stainless steel and nitrided alloy steel. The wear-resistance of the dry film on stainless steel increased with the rise of solidification temperature and reached the maximum value at 390°C; when the temperature continued to raise, the wear-resistance decreased due to the oxidation of MoS_2 and the decomposition of polyphenylene sulfide. The solidification temperature was appropriate at 370°C or 380°C when the substrate was nitriding alloy steel. Table 5.3 indicated that the wear-resistance of the film solidified at 380°C for 1.5h was higher when the substrate was the nitriding alloy steel.

Table 5.2　Effect of solidification temperature on the wear-resistance of dry film [9]

Solidification temperature /°C	Solidification time /h	Stainless steel		Nitriding alloy steel	
		Wear-resistance /(m/μm)	Average wear-resistance /(m/μm)	Wear-resistance /(m/μm)	Average wear-resistance /(m/μm)
360	2	290—314	302	254—342	303
370	1.6	284—394	335	307—422	360
380	1.5	294—413	345	314—449	362
390	1.5	309—493	386	The bonding was not strong	The bonding was not strong
400	1.5	277—375	336	—	—

Table 5.3　Effect of solidification time on the wear-resistance of dry film [9]

Solidification temperature/°C	Solidification time/h	Nitriding alloy steel	
		Wear-resistance/(m/μm)	Average wear-resistance/(m/μm)
380	0.5	203—349	296
	1.0	273—352	323
	1.5	314—449	362
390	0.5	284—310	299
	1.0	262—390	331
	1.5	The bonding was not strong	The bonding was not strong

For the requirements of reducing friction and extending life to the aero-engine, the researchers had developed bonded MoS_2 coating of model FM-510. The coating possessed high load bearing-capacity and excellent anti-wear behavior [10].

The friction and wear tests were carried out on an MRH-3 ring-on-block tester and Falex tester. The ring material was carbon steel with the diameter of 49.2mm and hardness of HRC 60; the block materials were 1Cr18Ni9Ti and 18CrMoWA steels. The friction-pairs were formed as follows: ① The ring was coated, and the block was uncoated; ② 1Cr18Ni9Ti block and ring were both coated; ③ 18CrMoWA block and ring were both coated. The surface roughness of the blocks was R_a=0.05μm.

Figure 5.31 shows the variation of friction coefficient of the FM-510 coating with load. The friction coefficient gradually decreased with the increase of load. The friction coefficient of the friction-pairs ① and ② decreased to 0.06 when the load increased to 1200N; the friction coefficient of friction-pair ③ decreased to 0.06 when the load increased to 2000N. When the load was further increased to 2900N, there was no significant change.

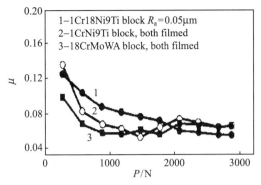

Fig.5.31   Variation of friction coefficient with load [10]

Figure 5.32 shows the variation of friction coefficient of frictional pair composed of coated ring and 1Cr18Ni9Ti coated block with time at load 2900N. When the sliding velocity was higher (0.77m/s), the friction coefficient was low; however, the life time was short showing the obvious increase of friction coefficient. On the other hand, at lower sliding velocity, although the friction coefficient was higher, the corresponding life time was longer. This result indicated that the bonded MoS₂ coating of model FM-510 was more suitable for the condition at low velocities.

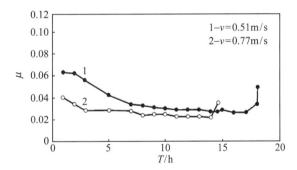

Fig.5.32   Variation of friction coefficient with time [10]

The bonded MoS₂ film is an effective measure to solve the lubrication problem in the harsh conditions, but MoS₂ lacks or even does not have protective ability in the corrosive environment and wet salt-fog medium; because its crystal surface can be oxidized to acidic product at lower temperature. The bonded film is usually required to have corrosion-resistance, which can be enhanced through phosphuration treatment on the substrate and adjusting the proportion of adhesive and solid lubricant [11].

Figure 5.33 shows the corrosion-resistance of the bonded solid lubrication film with and without phosphuration treatment on the low carbon steel substrate. The corrosion-resistance of the bonded film after phosphuration treatment of the substrate was significantly improved no matter whether the anti-corrosion additive of synthesis wax powders was added; because the phosphate film, which was insoluble in water and formed on the surface of phosphurized low carbon steel, played the role of preventing the corrosion of substrate.

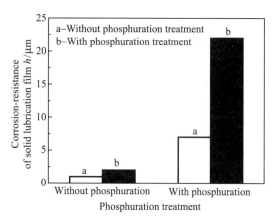

Fig.5.33   Effect of phosphuration treatment on the corrosion-resistance of the solid lubrication film [11]

Figure 5.34 shows the effect of the proportion of adhesive and solid lubricant on the corrosion-resistance of MoS₂ film. The bigger the proportion of adhesive/solid lubricant, the better the corrosion-resistance of the lubrication film.

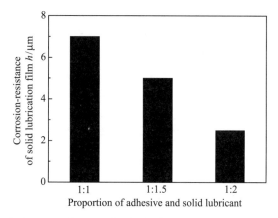

Fig.5.34   Effect of the proportion of adhesive/solid lubricant on the corrosion-resistance of MoS₂ film [11]

Some studies showed that epoxy resin bonded MoS₂ coating exhibits good fretting wear resistance. Preparation process, temperature, humidity and lubrication oil all have a certain impact on the fretting wear resistance of the bonded MoS₂ coating [12].

The substrate material also has significant effect on the fretting wear resistance of the bonded MoS₂ coating, as shown in Figure 5.35. The higher the hardness of substrate material, the longer the fretting wear life of the coating. Because when the load was applied on the coating surface, the coating and substrate simultaneously bore the load; the deformation of coating and substrate was different, thus there was different contact stress distribution at the interface, in consequence, the coating exhibited different fretting wear resistance during the reciprocating fretting process. In addition, in the process of fretting wear, with the continual spalling of the coating, part of the substrate surface directly contacted with the counterpart surface. The harder substrate surface was more wear-resistant and conducive to protect the coating in the substrate valley so that the coating life was greatly extended.

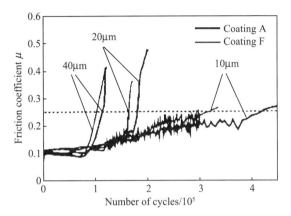

Fig.5.35   Effect of substrate material on the fretting wear life of the coating [12]

The fretting wear life can be improved when the thickness of the coating was increased, as shown in Figure 5.36. The thicker the coating, the longer the wear life. Under the lower displacement amplitude D ($<20\mu m$), the fretting wear life was longer at high loads, and the impact of load was weak, because under the lower displacement amplitude, the increasing load caused the enlargement of alternating stress, so that the spalling of the coating was intensified; meanwhile, the wear debris were difficult to overflow the contact area, and most of them could be pressed once again on the surface to supply the lubrication; as the result, the impact of the load variation was weakened.

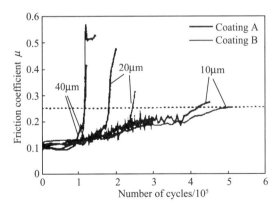

Fig.5.36   Effect of the coating thickness on the fretting wear life [12]

The fretting wear life of the MoS₂ coating can be also improved after solidification treatment at high temperature. Figure 5.37 shows the effect of solidification tempera-ture on the fretting wear life of the coating. The fretting wear life of the MoS₂ coating after solidification treatment at high temperature was obviously higher than that at room temperature; moreover, this difference was more obvious with the decrease of fretting displacement amplitude.

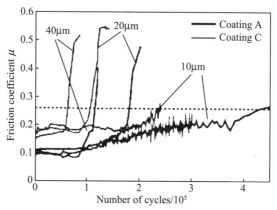

Fig.5.37   Effect of solidification temperature on the fretting wear life of the coating [12]

The above results indicated that the preparation process had significant effect on the fretting wear life of the MoS_2 coating, while the environment factors, such as humidity, temperature and lubrication oil can also affect it in different degree.

Figure 5.38 shows the effect of humidity on the fretting wear life of the coating. As the displacement amplitude increased, the sensitivity of wear life to the humidity variation increased. The higher the relative humidity, the lower the life, because the water vapor would infiltrate into the porosities of coating, the reciprocating fretting would cause the local high pressure to speed the wear and spalling of the MoS_2 coating, weaken the cohesive, and intensify the oxidation of MoS_2. All these factors would result in the shortened wear life. With the increase of displacement amplitude, more water vapor would enter into the contact area; therefore, the larger displacement amplitude could lead to the decrease of the wear life.

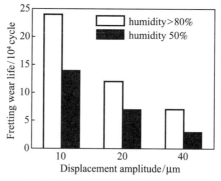

Fig.5.38   Effect of humidity on the fretting wear life of the coating [12]

Figure 5.39 shows the effect of temperature on the fretting wear life of the coating. When the temperature was between 10°C and 100°C, the higher the temperature, the higher the wear life of the coating; the wear life at 100°C was as three times long as that at 10°C. When the temperature was between 100°C and 300°C, the wear life decreased with the increasing of temperature. This might be due to that when the temperature was less than 100°C, the increasing of temperature led to decrease of humidity in the contact area so that the wear life was obviously improved. However, when the temperature was greater than 100°C, with the increase of temperature, the bonding strength of adhesive decreased; the oxidation damage of MoS₂ was intensified.

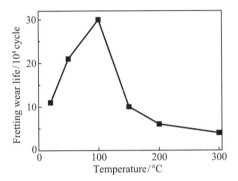

Fig.5.39 Effect of temperature on the fretting wear life of the coating [12]

The effect of lubrication oil on the fretting wear life of the MoS₂ coating is strongly dependent of the displacement amplitude and load as shown in Figs. 5.40 and 5.41. The lubrication oil hardly affected the wear life of coating at high load and low displacement amplitude (10μm) condition; while the wear life of coating sharply decreased at low load and high displacement amplitude (≥20μm) condition, and the wear life decreased with the increase of displacement amplitude. Because when the displacement amplitude was smaller or the load was larger, the contact surface was in the adhesion state, the lubrication oil was not easy to immerse the contact area. However, when the displacement amplitude was larger or the load was lower, the lubrication oil was easier to immerse the contact area, and a great deal of lubrication oil could fully immerse the interface of polymer adhesive fibers; as the oil was

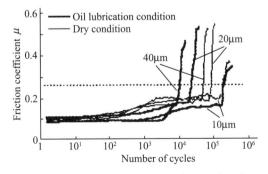

Fig.5.40 Comparison of the wear life of the coating under dry and oil lubrication conditions [12]

incompressible, when it bore the reciprocating alternating stress, it would rapidly tear down the adhesive fibers to accelerate the damage of the bonded coating; as a result, the wear life was greatly decreased.

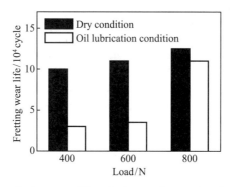

Fig.5.41 Comparison of the wear life of the coating under dry and oil lubrication conditions [12]

5.1.5   Inorganic Fullerene-like Nano MoS₂ Film

There is an obvious difference in the structure between the hollow fullerene-like nano MoS₂ and traditional layered MoS₂, as shown in Figure 5.42. The hollow fullerene-like nano MoS₂ has curved S-Mo-S planes; while the S-Mo-S planes of layered sputtering MoS₂ present basically the straight shape.

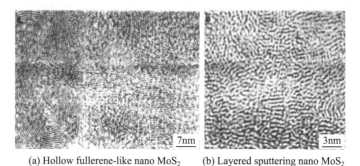

(a) Hollow fullerene-like nano MoS₂   (b) Layered sputtering nano MoS₂

Fig.5.42   HTEM images of hollow fullerene-like nano MoS₂ and layered nano MoS₂ [13]

Table 5.4 shows the tribological properties of hollow fullerene-like nano MoS₂ film and general 2H-MoS₂ film. Under the condition of relative humidity at 45%,

Table 5.4   Tribological properties of hollow fullerene-like nano MoS₂ film and general 2H-MoS₂ film [13]

	μ	$\omega/(\mathrm{mm^3N^{-1}mm^{-1}})$
Hollow fullerene-like nano MoS₂ film	0.008—0.01	1×10^{-11}
2H-MoS₂ film	0.1—0.3	3×10^{-1}

Note: μ is the friction coefficient under the condition of relative humidity at 45% at room temperature; ω is the wear volume.

the values of μ and ω of hollow fullerene-like nano MoS$_2$ film were extremely low, indicating that this kind nano MoS$_2$ was an excellent solid lubricant. The μ of hollow fullerene-like nano MoS$_2$ in the moisture air was only slightly higher than that (μ= 0. 006) in the dry nitrogen.

As well known, the tribological property of traditional MoS$_2$ solid lubricant is dependent of its typical layered structure; the layers are combined by Van der Waals bond and easy to be sheared, its lubrication effect is controlled by the shearing strength. However, the shearing mechanism can not control the hollow fullerene-like nano MoS$_2$; because the shearing can not take place between its internal and external dense layers. The hollow fullerene-like nano MoS$_2$ presents a round cage; its lubrication mode is rolling friction instead of sliding friction. In addition, the particles of nano MoS$_2$ are ultra-fine, they are sufficient to prevent the rough contact of lubrication surfaces.

Some studies showed that the friction-reducing and wear-resistance properties of the 2H-MoS$_2$ in dry argon and vacuum environment are better than those in the moist air. In the moist air environment, 2H-MoS$_2$ can be oxidized to MoO$_3$ and other more complex molybdenum oxides, such as Mo$_2$O$_3$, Mo$_2$O$_5$ and Mo$_3$O$_7$. In fact, the wear of 2H-MoS$_2$ crystal is related to its chemical reactions, which mainly occur at the position of its prismatic edge, where exist the dangling bonds susceptible to chemical reactions. While the crystal structure of hollow fullerene-like nano MoS$_2$ does not have such dangling bonds, so that it can prevent the MoS$_2$ from oxidation to molybdenum oxide to a great extent.

Hollow fullerene-like nano MoS$_2$ exhibits extremely low friction and wear in various environments, it is an excellent solid lubricant with potential application prospects.

5.2 MoS$_2$/metal Co-deposition

The sputtering MoS$_2$ film has been used in industries since the 1990s; it is suitable for lubricating the precision parts used in the special environment and has achieved good practical effect. However, as the target is subject to high energy ions bombardment during sputtering, the microstructures or compositions of the deposition film can be changed somewhat, compared with the target materials. The film performance can be improved to a certain extent when the sputtering parameters are controlled carefully, but the effect is limited.

Since the 1980s, in order to improve the performance of sputtering film, many researchers at home and abroad have been developing composite film composed of several materials with synergistic effect to replace the single film. For instance, MoS$_2$/metal co-sputtered film shows many advantages such as high bonding strength between the film and substrate, stable friction coefficient, long wear life, less debris, and good anti-wet behavior, compared with the single MoS$_2$ film. MoS$_2$/graphite sputtered film has excellent anti-wet performance; therefore, it can be used preferably in the higher relative humidity atmosphere. The sputtering MoS$_2$ film is loose and easy to be oxidized when it is stored in the air, while MoS$_2$/Ag co-deposited film is not oxidized and its wear-resistance can achieve as 24 times high as that of pure MoS$_2$ film. The friction coefficient of MoS$_2$/PTFE co-deposited film can be decreased from

0.2—0.3 to 0.1, its wear life increases by 10 times, compared with the single MoS₂ film at the same conditions. The co-deposited MoS₂/Au film can prevent its shearing deformation, increase the resistance to the crack initiation and propagation, and improve its wear life obviously.

5.2.1   MoS₂/Ni Composite Film

5.2.1.1   MoS₂/Ni co-sputtered composite film

As the MoS₂/Ni composite film is denser than MoS₂ film, it has less sensitivity to the friction load and sliding speed. No matter in the air, vacuum or argon, the life of co-sputtered film is obviously longer than that of simple sputtering film.

The friction coefficient of the MoS₂/Ni composite film deposited on the stainless steel decreased with the increase of load and tended to a stable state as shown in Figure 5.43. Therefore, MoS₂/Ni composite film is suitable to be used in the relatively high load ($> 3N$) condition.

The friction coefficient of MoS₂/Ni composite film decreased when the speed became larger, it decreased gradually from 0.132 at low speed to 0.051 when reaching the stable state as shown in Figure 5.44. It can be concluded that MoS₂/Ni composite film is also suitable to be used at high sliding speed ($>300r/min$) condition.

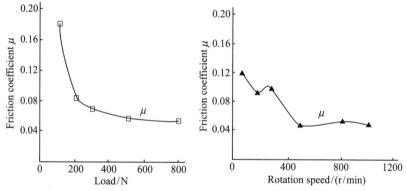

Fig.5.43   Variations of friction coefficient with load [14]

Fig.5.44   Variations of friction coefficient with sliding speed [14]

5.2.1.2   Ni/MoS₂ composite brush plating layer

Owing to the simplicity of technology and cheapness, the composite solid lubrication plating layer produced by the conventional plating method has been being attractive, and a lot of successes have been published. However, using the conventional plating method is difficult to get a composite plating layer with high content of MoS₂ and enough thickness, so that the friction-reducing function of MoS₂ is not developed satisfactorily. Thus its application is restricted.

Brush plating is similar to the conventional vat plating in principle, but the former shows some important characteristics, such as high density of current, high speed of deposition, and relative motion of the anode on the surface of work-piece during a plating process. A satisfactory composite plating layer can be obtained, if an appropriate solution and technology are adopted.

Some researchers used the brush plating method to prepare Ni/MoS$_2$ composite brush plating layer on the 1045 steel. The composite brush plating layer was studied comprehensively in the atmosphere and vacuum conditions. The results showed that the performance of the Ni/MoS$_2$ composite brush plating layer was closely related to the working conditions such as MoS$_2$ content, load, and sliding velocity.

1. Tribological performance of Ni/MoS$_2$ composite brush plating layer in the atmosphere

The friction and wear tests were conducted on a ball-on-disc testing machine. The upper sample was a 52100 steel ball with a radius of 6.35mm, hardness of HV 770, and the surface roughness of R_a=0.01μm; the lower sample was the 1045 steel disc with Ni/MoS$_2$ composite brush plating layer, with dimension of ϕ60mm×5mm, and hardness of HV 449. In the oil lubrication condition, a non-cyclic drop-feed lubrication with 20$^\#$ engine oil was adopted. In the beginning of tests, a running in period of 2min under 110N load was conducted for every new pair of specimens, and then the stepwise loading process was started by the rate of 100N for every 10s until scuffing. In the dry condition, a fixed load of 100N was used.

1) Tribological properties under oil lubrication condition

Figure 5.45 shows the variation of friction coefficient with the change of load for two composite plating layers with 75% and 11% MoS$_2$ at 0.97m/s sliding velocity. It indicates two main characteristics: ① The plating layer with lower MoS$_2$ content shows lower friction coefficient; ② when the load reaches 400—500N, a valley of friction coefficient is present, i.e., the friction coefficient decreases before it, and then increases. The explanations may be as follows.

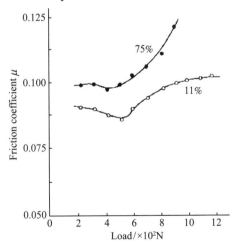

Fig.5.45   Variation of friction coefficient with load (v=0.97m/s)

(1) The variation of friction coefficients is deduced by the plastic deformation of a plating layer. The hardness of the plating layer with 75% MoS$_2$ is HV 50, whereas

that of the plating layer with 11% MoS$_2$ is HV 250, which is about 5 times that of the former. It is obvious that the plating layer with higher MoS$_2$ content can be plastically deformed more easily. This will consume greater deformation work and enhance the friction coefficients.

(2) In the experiment, 400—500N load may just make the rubbing pair being in the boundary lubrication state. Before that a partial elastohydrodynamic or mixed lubrication state is present, i.e., except oil film the boundary film can be formed as well, so when the load is increased, the boundary film will play a more important role, which can reduce the friction coefficient until the lubrication state becomes the boundary state completely. After that, the oil film loses its function, and the friction coefficient will be controlled mainly by the plastic deformation property of the plating layer.

The variation of friction coefficient with sliding velocity under 500N load for the composite plating layer with 40% MoS$_2$ is shown in Figure 5.46. It can be seen that the friction coefficient is decreased gradually with the increase of the speed. Such a phenomenon is still difficult to be explained now. Maybe it is related to the temperature effect, i.e., the sliding speed can increase frictional temperature more significantly than the load, and soften the Ni matrix of the composite plating layer; consequently, the frictional coefficient may be reduced.

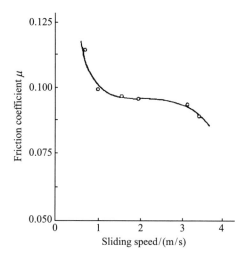

Fig.5.46 Variation of friction coefficient with sliding velocity

Figure 5.47 shows the relationship between the content of MoS$_2$ in a plating layer and a scuffing load. It can be found that a peak value of the scuffing load is always presented as 11% MoS$_2$ in spite of the sliding speed, and the value is corresponding to more than 8 times that of the pure Ni plating layer. This shows quite significant scuffing-resistance of MoS$_2$.

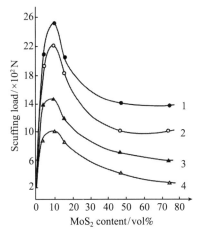

Fig.5.47 Variation of the scuffing load with MoS$_2$ content in the plating layer
($1-v=0.97$m/s; $2-v=1.26$m/s; $3-v=2.11$m/s; $4-v=3.59$m/s)

The appearance of the peak value can be explained by the increase of coverage rate of MoS$_2$ on rubbing surface before it until the whole surface is covered, and then the decrease of hardness of plating layer with higher content of MoS$_2$, which is easily worn and scuffed.

2) Tribological performance under dry friction condition

Figure 5.48 shows the relationship between MoS$_2$ content in the plating layer and friction coefficient in the condition of 100N load and at 0.67m/s speed. With the increase of MoS$_2$, the friction coefficient descends rapidly in the beginning until 30% MoS$_2$, and then turns to a steady value. The similar law is exhibited for different sliding speed.

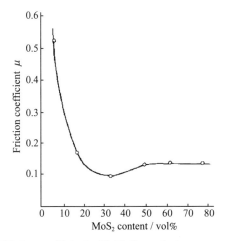

Fig.5.48 Variation of friction coefficient with MoS$_2$ content

The rapid drop of the friction coefficient for MoS$_2$ content $<30\%$ is obviously related with the coverage capacity of MoS$_2$ in the plating layer. When the content reaches the saturated state at 30%, its friction-reducing function cannot be further increased, even though the higher friction coefficient can be found due to the decrement of hardness of the composite plating layer (see Figure 5.45 and Figure 5.49). The saturated value, compared with that in oil lubricated condition, is shifted from 11% to 30%, which indicates that the lubrication oil helps the smearing effect of MoS$_2$ on the rubbing surface.

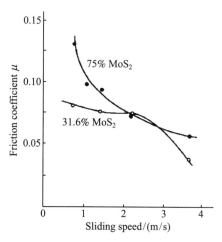

Fig.5.49   Variation of the friction coefficient with the sliding speed (P=100N)

The variation of the friction coefficient with the sliding speed for two composite plating layers with 31.6% and 75% MoS$_2$ under 100N load is shown in Figure 5.49. It is quite similar to Figure 5.46, i.e., the friction coefficient decreases with the increase of the sliding speed; especially, the amplitude of decrease for dry condition is even greater. Such a difference can be considered as an important experimental basis indicating the correctness of explanation about tribological behavior by temperature effect.

Figure 5.50 shows the relationship between MoS$_2$ content and wear life under 100N and at different sliding speeds. It can be found that a peak of wear life appeared at ∼30% MoS$_2$ content, which is corresponding to the saturated value of coverage by MoS$_2$. After that the wear life drops until 50%, and then a steady value is maintained.

Such behavior of MoS$_2$ is the result of its combined effects of friction-reducing and hardness decrement. When MoS$_2$ content is less than 30%, the increase of load bearing capacity of plating layer through friction-reduction is the dominant aspect. After 30%, the decrease of hardness and wear-resistance becomes dominant. As for the steady wear life for MoS$_2$ content higher than 50%, the reason is that the hardness of plating layer is already decreased to HV 120 and the thickness of plating layer at the center of wear track has been reduced quickly (only after 25 cycles) to ∼2μm by calculation. At this moment, the plastic flow stress of the plating layer is affected by its substrate, and the effect of hardness of the plating layer is already negligible.

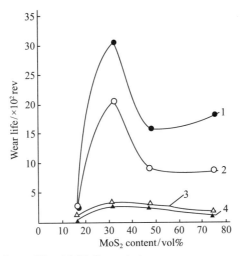

Fig.5.50   Variation of wear life with MoS₂ content
(P=100N; 1–v=0.67m/s, 2–v=0.97m/s, 3–v=2.11m/s, 4–v=3.59m/s)

2. Tribological performance of Ni/MoS₂ composite brush plating layer in vacuum

The friction and wear tests were conducted on a pin-on-disc testing machine in vacuum with a pressure of approximately $(6—9) \times 10^4$Pa. The material of the pins and discs was normalized 1045 steel with hardness HV 270. The diameter of the pin was 5mm, its roughness was R_a=0.02μm. Ni/MoS₂ composite plating layers with different MoS₂ contents and a thickness of 12μm were brush plated onto the surface of the disc.

Figure 5.51 shows the variation of friction coefficient with time for a composite plating layer with 31.6% MoS₂ content. The friction coefficient maintains a stable value from the very beginning, and is much lower than that in atmosphere. When the plating layer is worn away and scuffing occurs, the friction coefficient rises suddenly.

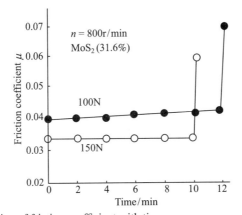

Fig.5.51   The variation of friction coefficient with time

Figures 5.52 and 5.53 show the variations in friction coefficient and wear life of the composite plating layer with MoS$_2$ content respectively. The friction coefficient only shows a lower stable value when the MoS$_2$ content is between 31.6% and 80%. The wear life of the composite plating layer is lengthened with an increase in MoS$_2$ content. It is quite different from the law of variation in atmosphere.

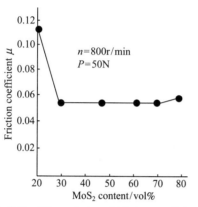

Fig.5.52　The variation of friction coefficient with MoS$_2$ content

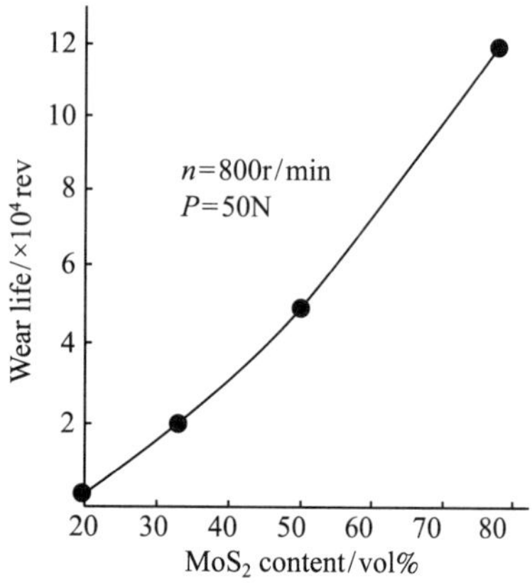

Fig.5.53　The variation of wear life with MoS$_2$ content

Figure 5.54 shows the variation in wear life of the composite plating layer with contact load. With increasing load, the wear life decreases rapidly. Figure 5.55 shows the relation between the friction coefficient of the composite plating layer with two different MoS$_2$ contents (31.6% and 78%) and load. The friction coefficient of the plating layer with high MoS$_2$ content decreases with increase of load more quickly than that with the lower content; finally it can reach approximately 0.014. This indicates that the composite plating layer with high MoS$_2$ content possesses excellent friction-reducing properties and wear life in vacuum under a moderate contact load.

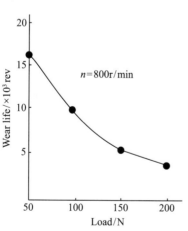

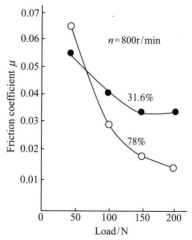

Fig.5.54　The variation of wear life with load

Fig.5.55　The variation of friction coefficient with load

Figure 5.56 shows the variation of friction coefficient of the Ni/MoS$_2$ (78%) composite plating layer with rotating speed. As a comparison the friction coefficient of a lead plating layer is also shown. The friction coefficient of the former is much lower than that of the latter, and is independent of rotating speed. In contrast, the friction coefficient of the lead plating layer increases with speed. At the same time, the friction-reducing property of the Ni/MoS$_2$ composite plating layer is almost unaffected by the environment; in vacuum it still shows good or even better solid lubrication properties. However, the friction-reducing property of the lead plating layer in vacuum is obviously poorer than that in atmosphere.

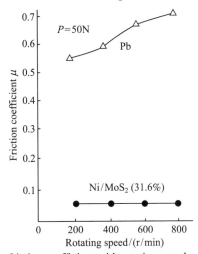

Fig.5.56 The variation in friction coefficient with rotating speed

There are a certain superiorities to prepare the metal/MoS$_2$ solid lubrication coating by brush plating technology: the volume fraction of MoS$_2$ in the plating layer can be adjusted in a range of 0—80%; whole or partial processing can be carried out for large and complex shape workpieces, which are difficult to be processed in the vacuum equipment. However, the MoS$_2$ is hard to avoid corrosive effects. Results have showed that the rare earth elements have an effect of catalytic reduction of SO$_2$; therefore, they can be used to prevent the oxidization of MoS$_2$.

The effects of rare earth Ce on the performances of Ni-Cu-P/MoS$_2$ plating layer were investigated on a ball-on-disc testing machine. The substrate material was 1045 steel disc, with dimensions ϕ55mm×5mm and surface roughness R_a =0.020μm. The brush plating layer with thickness of 10μm was made on its surface. The basic components in the plating layer and their mass fractions were Ni 59%, Cu 12%, P 4%, and MoS$_2$ 25%. Ce was added to the plating solution in the form of Ce(SO$_4$)$_2$. The specimens were stored in a sealed container with the temperature of 30°C and relative humidity of 80%. The bottom of the container is water, and the specimens were separated of water by a porous partition. The specimens were taken to be analyzed at predetermined time. The ball specimen was made from 52100 steel, with diameter 12.7mm and hardness HRC 64. The tests were performed under dry condition; the load of 10N was adopted.

Figure 5.57 shows the variation of friction coefficient of the Ni-Cu-P/MoS$_2$ plating layer with the Ce^{4+} content. Figure 5.58 shows the effect of holding time in the humid air on the friction coefficient of the plating layer without Ce^{4+}. Added Ce, the friction coefficient of the plating layer obviously decreased, whereas it was basically stable when the Ce^{4+} content was greater than 7.5g/L. With the increasing of holding time, the friction coefficient of the plating layer without Ce^{4+} increased continuously, especially fast when the layer was stored for one day; then the increasing rate obviously descended. The friction coefficient of the plating layer stored for 14 days was one time higher than that without storage.

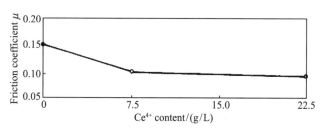

Fig.5.57 Variation of friction coefficient of the Ni-Cu-P/MoS$_2$ plating layer with the Ce^{4+} content

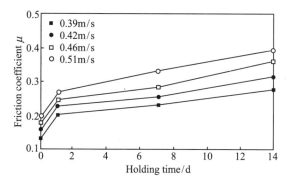

Fig.5.58 Effect of holding time on the friction coefficient of the plating layer without Ce^{4+}

Figure 5.59 shows the variations of the friction coefficients with holding time for two plating layers with 7.5g/L and 22.5g/L Ce^{4+} content. With the increasing of holding time, the friction coefficients of the two plating layers both increased slightly. This indicated that the humid air had little effect on the tribological performance of the plating layer with Ce^{4+}.

Above results proved that the rare earth Ce^{4+} could improve the friction-reducing performance of the Ni-Cu-P/MoS$_2$ plating layer, and also could effectively enhance the property of resisting humid air for plating solution.

When the friction coefficient sharply increases, the corresponding duration is regarded as the wear life of the plating layer. Figure 5.60 shows the variation of wear life with the Ce^{4+} content for the Ni-Cu-P/MoS$_2$ plating layer. When the Ce^{4+} content was 7.5g/L, the wear life of the plating layer was about 5 times higher than that of the plating layer without Ce^{4+}, whereas when the Ce^{4+} content continued to increase, it had little effect on the extension of wear life.

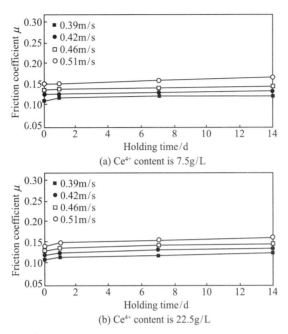

Fig.5.59   Variation of the friction coefficients with holding time for two plating layers with 7.5g/L and 22.5g/L Ce^{4+} content

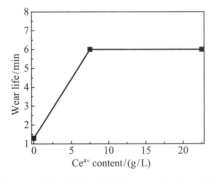

Fig.5.60   Variation of the wear life with the Ce^{4+} content for the Ni-Cu-P/MoS₂ plating layer

Under the action of water and oxygen, MoS₂ is easily subject to oxidation corrosion to cause the valence changes of elements. Figure 5.61 shows the XPS analysis results of S and Mo in the plating layer without Ce^{4+}. The S element in the plating layer stored for one day was corresponding to the S^{2-} position, which matched the S^{2-} of MoS₂ in the standard spectrum; the S^{6+} peak was present when the holding time was 6 days, and the S^{6+} position matched the S^{6+} of SO_4^{2-} in the standard spectrum, which indicated that the S^{2-} in the MoS$_2$ was oxidized. The Mo element in the plating layer stored for one day was corresponding to the Mo^{4+} of MoS$_2$ in the standard spectrum, whereas the Mo^{6+} peak appeared for the plating layer stored for 6 days, which indicated that the element Mo was also oxidized.

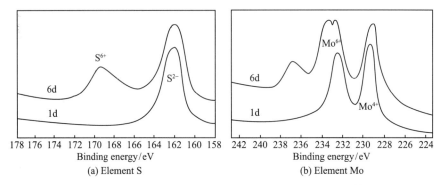

Fig.5.61   XPS analysis results of S and Mo in the plating layer without Ce^{4+}

Figure 5.62 shows the XPS analysis results of S and Mo in the plating layer with Ce^{4+}. The S element in the plating layer stored for one day was present in the form of S^{2-} and no new S peak appeared even stored for 15 days, which indicated that the sulfur in the MoS$_2$ co-existing with rare earth was not oxidized after experienced a longer period of time in the humid atmosphere. The valence of Mo in the plating layer stored for one day was still S^{4+}, and did not change even stored for 15 days.

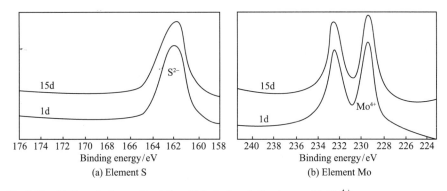

Fig.5.62   XPS analysis results of S and Mo in the plating layer with Ce^{4+}

Figure 5.63 shows the AES analysis results of the Ni-Cu-P/MoS$_2$ plating layer without Ce^{4+} before and after stored in the humid atmosphere. The oxygen in the plating layer stored for 6 days was obviously higher than that stored for one day, because a 0.12μm thick oxide film was generated on the surface of the plating layer.

In the humid atmosphere, MoS$_2$ is easy to be oxidized for the Ni-Cu-P/MoS$_2$ plating layer without Ce^{4+}. Mo^{4+}, S^{2-} are oxidized to Mo^{6+}, S^{6+}, respectively. As a result, the MoS$_2$ content in the plating layer is reduced; its self-lubricating property decreases, and the friction coefficient increases. When Ce^{4+} is added, the oxidization of MoS$_2$ in the humid atmosphere can be prevented effectively; the self-lubricating property of the plating layer will be stabilized, and the impact of humid air on the tribological performance of the plating layer will be reduced.

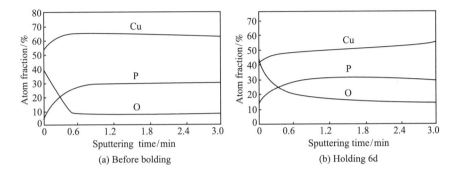

(a) Before bolding (b) Holding 6d

Fig.5.63 AES analysis results of the Ni-Cu-P/MoS$_2$ plating layer without Ce^{4+} before and after stored in the humid atmosphere

Figure 5.64 shows the XPS analysis results of Ce in the Ni-Cu-P/MoS$_2$ plating layer. There were two Ce peaks: one was corresponding to Ce^{3+} in the standard sample; another was the combined results of CeO$_2$ (881.6eV) and simple substance Ce (883.4eV). As the Ce was added to the plating solution in the form of Ce^{4+}, which was reduced to be electrodeposited during the brush plating process.

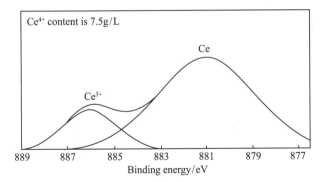

Fig.5.64 XPS analysis results of Ce in the Ni-Cu-P/MoS$_2$ plating layer

In order to further reveal the mechanism of Ce in the Ni-Cu-P/MoS$_2$ plating layer, the researchers removed the MoS$_2$ in the plating solution, cleaned it by distilled water and dried it; then it was analyzed by XPS after squash. It was found that there was only Ce on the MoS$_2$ surface, whereas no Ni existed, which indicated that Ce had a priority to be absorbed on the MoS$_2$ surface. This helps Ce to be deposited first on the MoS$_2$ surface to play the role of surface modification.

5.2.2   MoS₂/Ti Composite Film

MoS₂ film has been applied to the rubbing-pairs working in the vacuum environment; whereas it is hard to be used in the H_2O/O_2 atmosphere, where it will be oxidized to MoS_3 and its service life will be reduced. Many studies have proved that the MoS_2 film with Ti possesses excellent tribological performance, because Ti can play the role of preventing the oxidation of MoS_2 [15].

(1) Morphology.

Plasma immersion ion implantation and deposition (PIIID) was used to fabricate the MoS₂/Ti multilayer on the 2Cr13 substrate. The Ti layer was deposited by a pulse cathodic arc plasma source and the MoS_2 layer was obtained by a radio-frequency (RF) magnetron sputtering system. Different layers were prepared, and the processing parameters are shown in Table 5.5.

Table 5.5   Processing parameters of different layers [15]

Sample No.	Structure	Deposition time of MoS₂ layer/min	Deposition time of Ti layer/min	Total layer No.
S1	MoS₂/Ti	60	60	1
S2	MoS₂/Ti	60	60	2
S3	MoS₂/Ti	30	30	4
S4	MoS₂/Ti	20	20	4
S5	MoS₂/Ti	10	10	4
S6	MoS₂/Ti	15	15	8
S7	MoS₂	0	0	1

The SEM photograph of the cross-section of the MoS₂/Ti multilayer is shown in Figure 5.65. An obvious layered structure can be found in the multilayer. The total thickness of the multilayer is about 2μm, and the thickness ratio of MoS_2 to Ti layer is about 2:1. Since a high energy ion implantation was applied during the layer deposition, the interface between Ti and MoS_2 layer is not very clear.

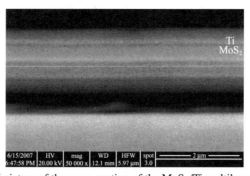

Fig.5.65   The SEM picture of the cross-section of the MoS₂/Ti multilayer [15]

(2) Tribological behavior.

Figures 5.66—5.68 show the influences of the total thickness of the multilayer, the MoS₂/Ti single layer thickness, and thickness ratio of MoS_2 to Ti layer on the friction behavior of the multilayer, respectively.

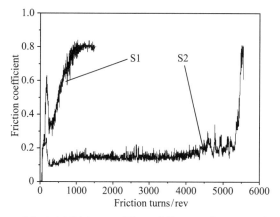

Fig.5.66   Influence of the total thickness of the multilayer on its wear resistance [15]

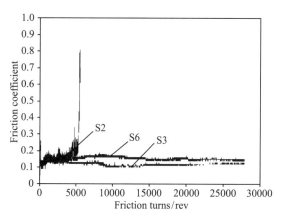

Fig.5.67   Influence of the MoS₂/Ti single layer thickness on tribological properties [15]

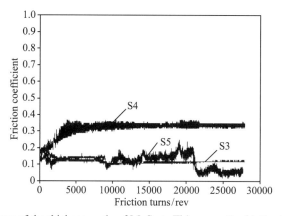

Fig.5.68   Influence of the thickness ratio of MoS₂ to Ti layer on the friction behavior [15]

It can be found from Figure 5.66 the cut-through number of sample S1 (thickness of 1μm) is about 300rev. However, when the total thickness of the multilayer is 2μm (sample S2), and the cut-through number exceeds 5000rev. Because of the larger total thickness, the load bearing capacity of sample S2 is larger than that of sample S1.

The multilayers in Figure 5.67 have the same total thickness of 2μm. It can be found that the cut-through number of sample S3 and S7 exceeds 28000rev, where that of sample S2 is only about 5000rev. In addition, sample S3 shows a lower friction coefficient. The crack propagation rate in the MoS₂/Ti layer can be decreased when the layer number is increased. Since all multilayers have the same total thickness of MoS₂ layers, and MoS₂ is the key anti-wear material in the film, the difference in wear resistance of different MoS₂/Ti multilayers should be caused by the crack propagation rate. Therefore, small MoS₂/Ti single layer thickness is benefit for the load bearing capacity and wear resistance of the multilayer.

The multilayers in Figure 5.68 have the same MoS₂/Ti layer thickness and to-tal thickness. The cut-through number of each multilayer exceeds 28000rev, which also proves that a small MoS₂/Ti single layer thickness can obtain a high wear resistance. However, the friction coefficient of each multilayer is different. For sample S3, the friction coefficient is 0.1 and keeps steady during the testing period. When the thickness ratio of MoS₂ to Ti layer is increased by one time (sample S4), the friction coefficient raises to 0.35. When the depth ratio is increased by four times (sample S4), the friction coefficient also decreases to 0.1, but it is not steady. These results reveal that the thickness ratio of 2:1 is beneficial for acquiring a stable MoS₂/Ti multilayer.

5.2.3   MoS₂/Au Co-sputtered Film

MoS₂/Au co-sputtered film possesses excellent tribological properties, such as high bonding strength with the substrate, stable friction coefficient, long wear life, less wear debris and good anti-wet performance [16]. Co-sputtered Au-MoS₂ coatings with a much higher range of metal contents up to (95at%) have shown surprisingly good performance at low contact stresses (as low as 0.1MPa).

The tribological test of Au-MoS₂ coatings was conducted using a reciprocating linear wear test apparatus. Figure 5.69 shows a plot of the coefficient of friction versus

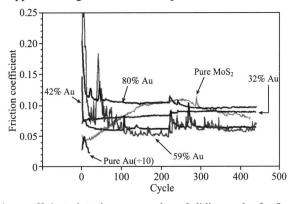

Fig.5.69 Friction coefficient plotted versus number of sliding cycles for five of the sputtered coatings in this study [16]

(The lens-on-disk apparatus was used at contact stress $S_m \approx 8.5$MPa. Data are shown for pure MoS₂, pure Au, and co-sputtered Au–MoS₂ coatings with 32at%, 42at%, 59at% and 80at% Au. The lowest COF values were for intermediate values of Au content, 42at% and 59at%.)

number of cycles for coatings with a range of Au contents, including pure Au and pure MoS_2. The coatings' endurances cannot be compared because none failed by the end of the test at 440 cycles, except for pure Au, which failed at 20 cycles. However, the relative performance of the coatings may be evaluated in terms of their average μ achieved after run-in. The coatings with 42at% and 59at% Au exhibited values of about 0.06 and 0.05, respectively. The coatings with 32at% and 80at% Au showed higher values (i.e., about 0.09 and 0.10, respectively). These results indicate that there is an intermediate Au content region of optimum performance.

5.3 MoS₂/ Metal Compound Composite Film

5.3.1 MoS₂/TiN Composite Film

The fretting tests were carried out on a SRV testing machine. A ball-on-disc contact was used. The upper specimen was 52100 steel ball with diameter of 10mm. The roughness of ball surface was R_a=0.08mm. The lower disc specimens were made from 1045 steel with dimensions of ϕ24mm×7.8mm. They were heated at 860°C for 20min, then water-quenched and tempered at 600°C for 30min. The roughness of their surface was also R_a=0.08mm.

Three kinds of coating were made on the cleaned disc specimens: ① ion-plated TiN with thickness of 2μm; ② magnetron-sputtered MoS_2 with thickness of 1μm; ③ at first, TiN was deposited with thickness of 1.8mm by ion-plating and then a MoS_2 coating was sputtered onto TiN coating with thickness of 1.5μm.

The main parameters of fretting tests were as follows: temperature, 25°C; relative humidity, 60%±10%; slip amplitude, 50μm to 100μm; normal load, 10N to 20N; frequency, 10Hz; number of cycles, up to 15000 cycles.

5.3.1.1 Friction coefficient

The variations of friction coefficient of TiN, MoS_2 and their composite coatings with the number of cycles (normal load 20N, slip amplitude 100μm, frequency 10Hz) are shown in Figure 5.70. It can be seen that the friction coefficient of TiN coating rapidly increased to 0.3 at the beginning and slowly reached the maximum value 0.33 until 6000 cycles, and then it slowly diminished to the steady state of 0.32. Compared with TiN coating, the friction coefficient of MoS_2 varied in a wide range of values. It

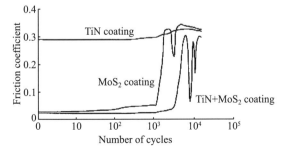

Fig.5.70 Variation of friction coefficient of three kinds of coating with number of cycles

showed that the friction coefficient kept a very low level of 0.055 before 1200 cycles, and then suddenly increased to 0.33 at 2700 cycles. However, at about 3000 cycles, the friction coefficient showed a sharp drop to 0.23, and then slowly decreased from 0.34 to a steady state of 0.32. A great difference was observed in the variation of friction coefficient for TiN + MoS₂ composite coating. The friction coefficient was very low (<0.04) during the incubation period (<3500 cycles); then it increased sharply to 0.30 until 6000 cycles. During the following 1800 cycles, the friction coefficient diminished down to 0.068 and then it rose rapidly to 0.28 after 1200 cycles. Before it reached the steady state of 0.32, the value displayed a sharp drop from 0.28 to 0.168.

5.3.1.2   Wear behavior

The variation of wear volumes of TiN, MoS₂, TiN + MoS₂ composite coatings and 1045 steel substrate with number of cycles under the same testing condition (normal load 10N, slip amplitude 100mm, frequency 10Hz and 15000 fretting cycles) are displayed in Figure 5.71. It indicated that the fretting wear-resistance of TiN+MoS₂ composite coating was the best and the anti-wear characteristic of TiN coating was superior to that of the single MoS₂ coating. The fretting wear resistance of 1045 steel was the worst among four kinds of materials. That means an appropriate surface coating is very effective for preventing the fretting wear, especially the duplex treatment, which can mutually supplement the advantages and disadvantages of the single coatings.

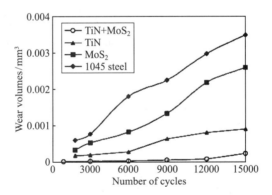

Fig.5.71   Variation of wear volumes of coatings and the substrate with number of cycles

5.3.1.3   Wear mechanism

MoS₂ is a widely used solid lubricant characterized by its hexagonal structure and low shearing strength. However, its low hardness usually results in poor wear resistance and short endurance, especially when its substrate is not strong enough. The morphology of wear debris (Figure 5.72) indicated that MoS₂ coating detached in plate-like shape. These MoS₂ debris was easily trapped in the contact area as a third body and could well play the lubricating role, which resulted in the sudden decrease of friction coefficient. The following rapid increase of friction coefficient probably could be explained by the contact of upper specimen with the substrate when the MoS₂ debris was not enough in the contact area. At last the friction coefficient approached to the steady state when the MoS$_2$ debris was squeezed out completely. One important phenomenon should be noticed, i.e., the MoS$_2$ and its oxides like MoO$_3$ could be detected as a boundary film by the EDX and XPS analyses as shown in Figure 5.73 and Figure 5.74 (d). That means, in the fretting process the MoS$_2$ coating could be not only transferred to the wear debris, but also to a boundary film. This thin film might be helpful to lengthen the role of MoS$_2$ coating as well.

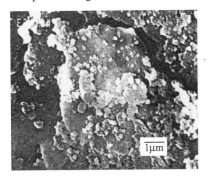

Fig.5.72   Debris morphology of MoS$_2$ coating

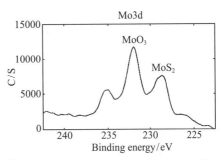

Fig.5.73   XPS analyses of the boundary film on the wear scar of MoS$_2$ coating

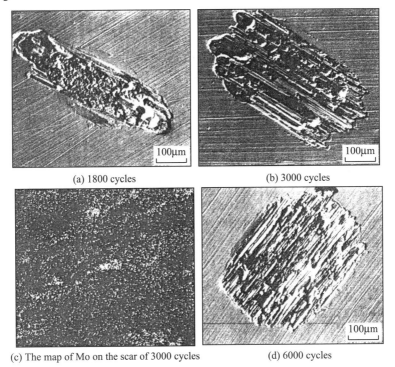

(a) 1800 cycles

(b) 3000 cycles

(c) The map of Mo on the scar of 3000 cycles

(d) 6000 cycles

Fig.5.74   Fretting wear scar morphologies of MoS$_2$ coating

As expected, the composite coating comprehensively played the role of two single coatings and showed desirable synergetic effect, i.e., the hard TiN coating offered a very strong support to the MoS$_2$ coating, so that the latter could fully play the role of solid lubricant and showed much longer endurance. In contrast, the MoS$_2$ coating reduced the friction coefficient of TiN coating and also lengthened its fretting life. The fretting wear mechanism of composite coating was not new, compared with that of TiN and MoS$_2$ coatings separately, but the damage process was considerably postponed, and more fluctuations of friction coefficient were present before the complete damage of the composite coating. At 1800 cycles (3min) only small portion of MoS$_2$ coating detached (Figure 5.74 (a), Figure 5.75 and Figure 5.76). At 3000 cycles (5min), most part of MoS$_2$ coating still remained (Figure 5.74 (b)). Only after 6000 cycles (10min), the MoS$_2$ coating fully detached (Figure 5.74 (c)), which was correspondent to the first peak of friction coefficient. The damage of TiN coating was also postponed to 9000 cycles (15min) (Figure 5.74 (d)). The second fall of friction coefficient was slighter than the previous one, which might be due to the combined effect of TiN, MoS$_2$ and substrate wear debris in the contact area.

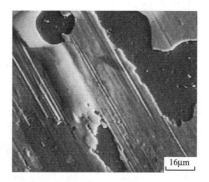

Fig.5.75   Debris detachment of MoS$_2$ coating

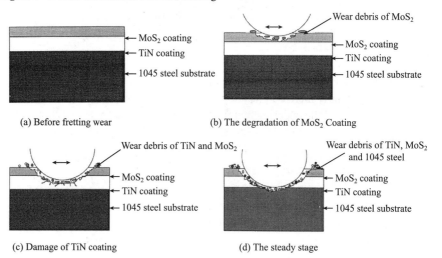

(a) Before fretting wear

(b) The degradation of MoS$_2$ Coating

(c) Damage of TiN coating

(d) The steady stage

Fig.5.76   The schematic diagram of fretting damage mechanism of TiN + MoS$_2$ composite coating

5.3.2 MoS₂/Pb₂O₃ Composite Film

Figure 5.77 shows the variation of friction coefficients of MoS_2/Pb_2O_3 composite films in the ultra-high vacuum and atmospheric environments. In the ultra-high vacuum condition, the largest and average friction coefficients of MoS_2/Pb_2O_3 were 0.13 and 0.07, respectively; they diminished to about 1/3 those in atmosphere; meanwhile, the fluctuation of average friction coefficient also decreased obviously. This indicated that the tribological property of the MoS_2/Pb_2O_3 composite film in the ultra-high vacuum was superior to that in the atmosphere. The wear life of the MoS_2/Pb_2O_3 film with thickness of 0.8μm has reached 3.7×10^6rev in atmosphere; thus the wear life in the ultra-high vacuum will be longer.

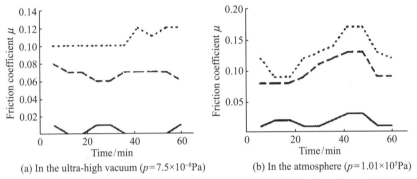

(a) In the ultra-high vacuum ($p=7.5\times10^{-8}$Pa)   (b) In the atmosphere ($p=1.01\times10^5$Pa)

Fig.5.77 Variations of friction coefficients of MoS_2/Pb_2O_3 composite film [17]

(· · · largest friction coefficient; - - -average friction coefficient;—fluctuation of average friction coefficient)

5.3.3 MoS₂/LaF₃ Composite Film

Figures 5.78 and 5.79 are TEM and SEM morphologies of MoS_2/LaF_3 co-sputtered film and pure MoS_2 respectively [18]. The branch-like protrusions of the co-sputtered film were coarser, and its columnar structure was denser, compared with those of pure MoS_2 film. This special structure prevented the diffusion of vapor into the MoS_2/LaF_3 film and significantly increased its anti-wet performance.

Table 5.6 shows the experimental results of the MoS_2/LaF_3 co-sputtered film and MoS_2 sputtered film after storage in air and vacuum with 100% RH. The MoS_2/LaF_3 co-sputtered film possessed long wear life in air and vacuum; the wear life became even longer after storage in humid environment for a certain period of time. Compared with MoS_2, MoS_2/LaF_3 film had higher wear life, and the variation of wear life was smaller with the increase of storage time. The friction coefficients of the MoS_2/LaF_3 co-sputtered film basically did not change with the variation of storage time and always kept at about 0.10. When a small amount of LaF_3 was added, the structure of co-sputtered film would be dense, which inhibited the oxidization of MoS_2 to a certain extent, prevented the corrosion of substrate, and stabilized the film structure. In addition, the MoS_2/LaF_3 co-sputtered film also possessed excellent anti-wet performance.

(a)   (b)

Fig.5.78   TEM and SEM morphologies of MoS₂/LaF₃ co-sputtered film [18]

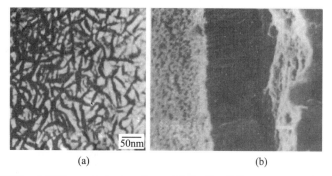

(a)   (b)

Fig.5.79   TEM and SEM morphologies of pure MoS₂ film [18]

Table 5.6   Variation of tribological properties with storage time for MoS₂/LaF₃ co-sputtered film and MoS₂ sputtered film [18]

Stored time/day			0	5	10	15
MoS₂-LaF₃ co-sputtered film	Wear life/min	Air	145	173	171	135
		Vacuum	150	180	140	95
	Friction coefficient	Air	0.09	0.10	0.10	0.10
		Vacuum	0.09	0.10	0.10	0.10
MoS₂ sputtered film	Wear life/min	Air	135	128	111	49
		Vacuum	105	98	—	41
	Friction coefficient	Air	0.09	0.08	0.08	0.11
		Vacuum	0.08	0.07	—	0.11

The researches have shown that LaF₃ has a great effect on the tribological properties of MoS₂ dry film bonded with phenolic-epoxy. When LaF₃ is added, the wear life of dry film will be extended, however LaF₃ has no obvious impact on the friction coefficient of dry film. Furthermore, LaF₃ can also reduce the oxidation trend of MoS₂ in the dry film during the friction process.

5.3.4   MoS₂/FeS Multilayer Film

The magnetron sputtering + low temperature ion sulfurizing composite technology was utilized to prepare MoS₂/FeS multilayer film on the 1045 steel heat–treated to hardness of HRC 55. Mo/Fe multilayer film was made in the magnetron sputtering equipment of model GDM-450BN. The dimension of Fe target (99.9% purity) and Mo (99.9% purity) target was 124mm×254mm×2mm and 124mm×254mm×10mm, respectively. The distance between samples and targets was about 70 mm. The Mo/Fe multilayer film was designed with 4 layers composed of alternative deposition (thickness of each layer: 500nm), the total thickness of the multilayer film was 2μm. The Mo/Fe multilayer film was then treated by low temperature ion sulfuration for 2h to form MoS$_2$/FeS multilayer film.

5.3.4.1   Characterization

Figure 5.80 shows the surface, cross-section morphologies and elements distribution of the MoS$_2$/FeS multilayer film. The surface was quite smooth, and plenty of spherical particles were present. On the cross-section, MoS$_2$/FeS multilayer film presented a ribbon white layer with homogeneous thickness about 2μm.

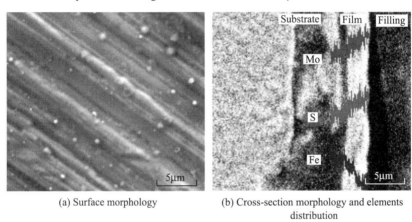

(a) Surface morphology   (b) Cross-section morphology and elements distribution

Fig.5.80   Surface, cross-section morphologies and element distribution of the MoS$_2$/FeS multilayer film

Figure 5.81 shows the XRD pattern of MoS$_2$/FeS multilayer film. The diffraction peaks are mainly composed of Fe, Mo, FeS and MoS$_2$. Mo/Fe multilayer film possessed plenty of grain and layer boundaries, which were helpful to the diffusion

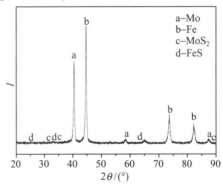

Fig.5.81   XRD pattern of MoS$_2$/FeS multilayer film

of sulfur into the Mo/Fe film during sulfurizing process. Figure 5.82 shows the load-displacement curves of MoS₂/FeS multilayer film.

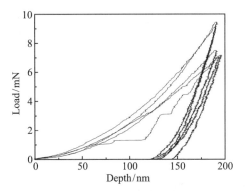

Fig.5.82   Load-displacement curves of MoS₂/FeS multilayer film

5.3.4.2   Tribological properties of MoS₂/FeS multilayer film

Figure 5.83 shows the curves of tribological properties of the MoS₂/FeS multilayer film, FeS film, and 1045 steel under dry condition. The friction coefficient of the MoS₂/FeS multilayer film was quite low and stable; during the whole test, it was obviously lower than that of the FeS film and 1045 steel. As the MoS₂/FeS multilayer film was soft, at the beginning of the test, its wear scar depth was slightly higher than that of FeS film and 1045 steel. After 8min, the MoS₂/FeS multilayer film gradually played the role of solid lubrication, and its depth of wear scar was obviously lower than that of FeS film and 1045 steel.

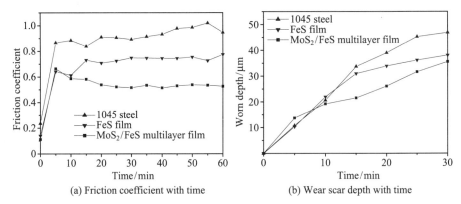

(a) Friction coefficient with time   (b) Wear scar depth with time

Fig.5.83   Variation curves of tribological properties of the MoS₂/FeS multilayer film, FeS film, and 1045 steel under dry condition

Figure 5.84 shows the worn morphologies and compositions of the 1045 steel, FeS film and MoS₂/FeS multilayer film under dry condition after wear test of 60min. Many wide and deep furrows and severe plastic deformation were present; the mechanism was abrasive wear and adhesive wear. The wear loss of the FeS film was relatively small compared with that of the 1045 steel, but some cracks on the film surface were present. The worn surface of MoS$_2$/FeS multilayer film was smooth and no obvious wear scar was found. Figure 5.84 (d) shows that there was almost no sulfur element on the worn surface of FeS film, which means, the FeS film had been entirely destroyed. While a certain S and Mo elements could still be detected on the worn surface of MoS$_2$/FeS multilayer film as shown in Figure 5.84 (f).

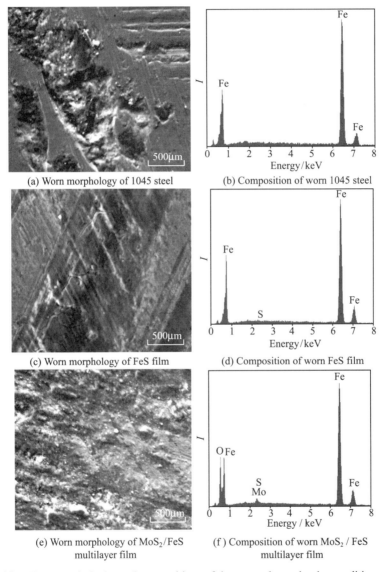

(a) Worn morphology of 1045 steel

(b) Composition of worn 1045 steel

(c) Worn morphology of FeS film

(d) Composition of worn FeS film

(e) Worn morphology of MoS$_2$/FeS multilayer film

(f) Composition of worn MoS$_2$ / FeS multilayer film

Fig.5.84   Worn morphologies and compositions of three samples under dry condition

Figure 5.85 shows the tribological property curves of MoS$_2$/FeS multilayer film, FeS film, and 1045 steel under oil lubrication condition. It was found that the friction coefficient of the MoS$_2$/FeS multilayer film deceased rapidly at the early stage of wear test before attaining a steady state; it showed a very low value throughout the entire test, and was always lower than that of the FeS film and 1045 steel. The depths of wear scars of the three samples all increased with the increase of time, but the increase rate of the MoS$_2$/FeS multilayer film was very small. The wear scar depth of the multilayer film was always less than that of the FeS film and 1045 steel.

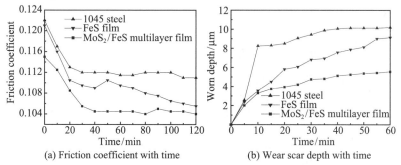

(a) Friction coefficient with time   (b) Wear scar depth with time

Fig.5.85   Variation curves of tribological properties of MoS$_2$/FeS multilayer film, FeS film, and 1045 steel under oil lubrication condition

Figure 5.86 shows the worn morphologies and compositions of the 1045 steel, FeS film and MoS$_2$/FeS multilayer film under oil lubrication condition after wear test of 120min.

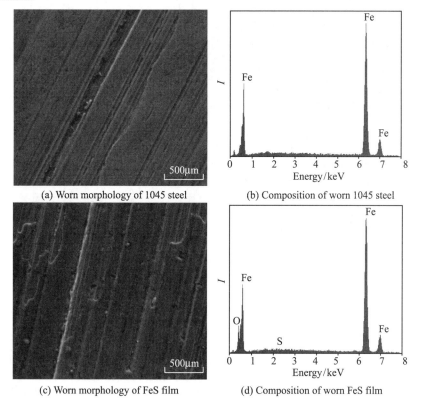

(a) Worn morphology of 1045 steel   (b) Composition of worn 1045 steel

(c) Worn morphology of FeS film   (d) Composition of worn FeS film

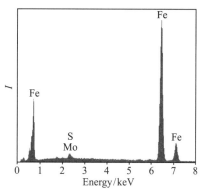

(e) Worn morphology of MoS$_2$/FeS multilayer film   (f) Composition of worn MoS$_2$/FeS multilayer film

Fig.5.86   Worn morphologies and compositions of the three samples under oil lubrication condition

5.3.4.3   Lubrication mechanism of the MoS$_2$/FeS multilayer film

Both FeS and MoS$_2$ possess the close-packed hexagonal crystalline structure; they can easily slip along the close-packed plane. Therefore, the MoS$_2$/FeS multilayer film possesses low friction coefficient. At the early stage of friction and wear test, the MoS$_2$/FeS multilayer film is subject to the plastic deformation forming a plastic flow layer on the wear surface; meanwhile, the wear debris of film can transfer to the counterpart surface to form a transfer film. Consequently, the friction coefficient and surface damage can be reduced. As the single FeS and MoS$_2$ film are soft, their wear-resistance is poor. However, the wear resistance of MoS$_2$/FeS multilayer film can be increased obviously because the surface hardness is increased due to the increment of dislocation density between layers. Moreover, in the process of magnetron sputtering and sulfurizing, the ion bombardment can also increase the hardness of the MoS$_2$/FeS multilayer film. Meanwhile, after bombardment, the compressive residual stress is generated on the film surface, which can reduce the tendency of crack generation. In addition, the FeS and MoS$_2$ can be oxidized to form stable oxides, which are hard and can protect the substrate metal to reduce the occurrence of surface adhesion. These characteristics determine the excellent friction-reducing and wear-resistance properties of the MoS$_2$/FeS multilayer film.

5.4   MoS$_2$/graphite Sputtered Coating

Figures 5.87 and 5.88 are the variations of friction coefficient and wear rate with load for the MoS$_2$/graphite sputtered coating in vacuum [19]. The friction coefficient was greatly affected by the load variation and diminished with the increase of load. The wear rate was increased slightly with the increase of load before 10N, when the load was more than 10N, the wear rate increased rapidly.

Figures 5.89 and 5.90 show the worn morphologies of the MoS$_2$/graphite sputtered coating at loads of 8N and 16N in vacuum. In the low load condition, the coating surface was slightly polished (see Figure 5.89 (a)) without obvious plastic deformation; some small cracks were present only in local area (see Figure 5.89 (b)). This indicated that the main wear mechanism of the coating was fatigue wear in the low load condition. Under the control of fatigue wear, the increment of load in a certain range can not significantly increase the wear rate, while in the high load condition, obvious plastic deformation took place on the worn surface, the deformed material piled up on both sides of the wear track (see Figure 5.90 (a)). The cracks caused by deformation were developed, meanwhile, the flake-like debris were formed (see Figure 5.90 (b)), which caused the increase of wear rate obviously in the high load condition. Step-like morphologies were generated on the worn surface due to the material flow caused by plastic deformation (see Figure 5.90 (b)). This phenomenon resulted in the roughening of the contact surface and increasing the ploughing action of the counterpart on the coating, which could also increase the wear rate.

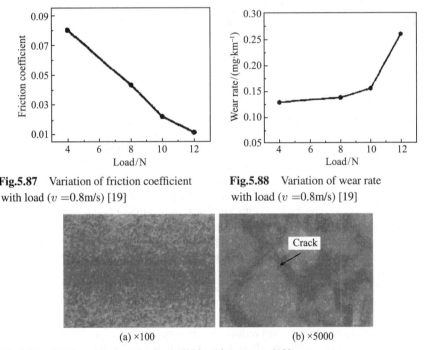

Fig.5.87 Variation of friction coefficient with load ($v =0.8$m/s) [19]

Fig.5.88 Variation of wear rate with load ($v =0.8$m/s) [19]

(a) ×100   (b) ×5000

Fig.5.89 SEM worn morphologies at 8N load in vacuum [19]

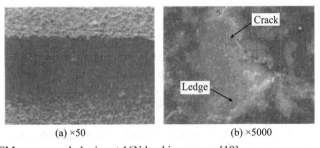

(a) ×50   (b) ×5000

Fig.5.90 SEM worn morphologies at 16N load in vacuum [19]

When the solid lubricants are selected appropriately, their synergistic effect can be used to obtain the coatings with excellent performance. Some studies have shown that the bonded coating made by MoS_2 and graphite can sustain a certain degree of high temperature and possesses good friction-reduction behavior.

As a lubricant, MoS_2 and graphite possess excellent friction-reducing performance, whereas their wear-resistance properties are poor. When the PTFE is used as a primer, it can not only increase the bonding strength between the coating and substrate, but also improve the lubrication performance of the coating due to the combined effect.

References

1. Wang J H, Yang J. Growth characteristics of MoS_2 coatings prepared by magnetron sputtering [J]. Lubrication Engineering, 2005, 6: 12–14.
2. Shao H H, Chen W. Deposition of MoS_2 films by radio frequency magnetron sputtering [J]. Lubrication Engineering, 2007, 32(12): 43–46.
3. Jiang X F, Liu Q C, Wang H B. Technology and application of multi-arc ion plating [J]. Journal of Chongqing University (Natural Science Edition), 2006, 29(10): 55–57.
4. Wang H D, Xu B S, Liu J J, et al. Microstructures and tribological properties on the composite MoS_2 films prepared by a novel two-step method [J]. Materials Chemistry and Physics, 2005, 91(2–3): 494–499.
5. Wang H D, Xu B S, Liu J J, et al. Wear-resistance and anti-scuffing of multi-arc ion plating molybdenum films [J]. Transactions of Nonferrous Metals Society of China, 2004, 14(Special 2): 338–343.
6. Wang H D, Xu B S, Liu J J, et al. Characterization and anti-friction on the solid lubrication MoS_2 film prepared by chemical reaction technique [J]. Science & Technology of Advanced Materials, 2005, 6(5): 535–539.
7. Wang H D, Xu B S, Liu J J, et al. Molybdenum disulfide coating deposited by hybrid treatment and its friction-reduction performance [J]. Surface & Coatings Technology, 2007, 201(15): 6719–6722.
8. Guan Y H, Xu Y, Zheng Z Y, et al. Microstructure and tribological properties of nano- and submicron-structured MoS_2 coating deposited by plasma spraying [J]. Transactions of the China Welding Institution, 2005, 26(12): 51–54.
9. Wang M L, Su F L, Li T S, et al. A study on PPS-2 dry film lubricant. Proceedings of the Third International Symposium on Tribology, 1983: 19–25.
10. Zheng Y H, Li J S, Wang M L. Friction characteristic of FM-510 bonded solid film [J]. Tribology, 1998, 18(4): 373–376.
11. Ye Y P, Chen J M, Zhou H D. A study on the corrosion-resisting properties of resin bonded MoS_2 solid lubrication films [J]. China Surface Engineering, 2001, (2): 26–28.
12. Xu J, Zhu M H, Liu H W, et al. Influence of relative humidity, temperature, oil lubrication on the fretting wear life of bonded solid lubricant coating [J]. Materials for Mechanical Engineering, 2003, 27(9): 21–23.
13. Zhang W Z. MoS_2 and WS_2 nanoparticles with hollow fullerene-like [J]. China Molybdenum Industry, 2002, 26(4): 18–20.
14. Yao G W, Li C S, Zhang Y. Preparation and tribology performance of magnetron sputtering MoS_2/Ni composite films [J]. Materials for Mechanical Engineering, 2007, 31(4): 52–54.
15. Wang L P, Zhao S W, Xie Z W, et al. MoS_2/Ti multilayer deposited on 2Cr13 substrate by PIIID [J]. Nuclear Instruments and Methods in Physics Research B, 2008, 266: 730–733.

16. Lince J R, Kim H I, Adams P M, et al. Nanostructural, electrical, and tribological properties of composite Au–MoS$_2$ coatings [J]. Thin Solid Films, 2009, 517: 5516–5522.
17. Chen T, Yan C L, Zhao Y P, et al. The properties of MoS$_2$ + Sb$_2$O$_3$ anti-cold welding thin film by magnetron sputtering [J]. Vacuum & Cryogenics, 2001, 7(3): 139–143.
18. Wang J A, Yu D Y, Ou Yang J L. Tribological properties of MoS2-LaF3 cosputtered films after stored in moist air [J]. Materials Science Progress, 1992, 6(3): 245–249.

Chapter 6

Micron-nano WS$_2$ Solid Lubrication Film

At present, MoS$_2$ as a solid lubrication film has widely been used in the industries, while there are still less researches and applications on WS$_2$ being reported. Compared with MoS$_2$, WS$_2$ possesses better anti-oxidation performance and is adaptable to a wider range of temperature; meanwhile, its oxidation product WO$_3$ has a lower friction coefficient than MoO$_3$. Therefore, it is worthy to pay more attention to WS$_2$ solid lubrication film.

WS$_2$ is an important lubrication material. It is not only applicable to the conventional conditions, but also for the harsh conditions, such as high temperature, high pressure, high vacuum, high load, irradiation, and corrosive media, etc. Therefore, in today's development of ever-changing technology, it has increasingly widespread usage. At present, there are a lot of methods to prepare WS$_2$ films or coatings, but their microstructure and tribological properties are quite different.

6.1   WS$_2$ Film

In order to extend the application range of solid lubricant WS$_2$, a new method—two step method was adopted to prepare WS$_2$ film; namely, W film was first fabricated on the AISI 1045 steel by magnetic sputtering method, and then treated by low temperature ion sulfuration to obtain WS$_2$ composite film [1,2]. Better tribological properties of the WS$_2$ composite film were achieved.

The substrate materials were 1045 and M2 steels. The specimens used for T-11 ball-on-disc tester were 1045 discs with dimensions of ϕ25.4mm×6mm and hardness HRC 55; The specimens used for MMS-1G high temperature tester were cylindrical pins with dimensions of ϕ14mm×40.3mm and hardness HRC 63—64. The W films with thickness of 2μm and 1μm were prepared on the 1045 and M2 steels by magnetron sputtering, and then treated by low temperature ion sulfuration to form WS$_2$ synthetic films.

The W target had a purity of 99.99%, with the dimension of 124mm×254mm×2mm. When the vacuum degree in the deposition chamber reached 3×10^{-3}Pa, Ar was injected. The pressure in the chamber was controlled at about 1.2Pa, and a constant current of 1.5A was set for deposition. The deposition time on the 1045 steel and M2 steel are 15min and 7.5min, respectively.

The W films were then treated by low temperature ion sulfuration for 2 hours.

6.1.1   Characterizations of the Synthetic WS$_2$ Film

Figure 6.1 shows the surface morphology of the synthetic WS$_2$ film. It was found that there were many spherical particles on the surface with dimension below 100nm.

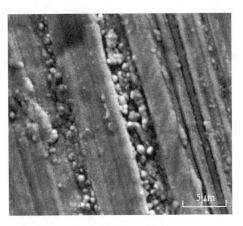

Fig.6.1   Surface morphology of the synthetic WS$_2$ film

Figure 6.2 shows the XRD patterns of the WS$_2$ film. The diffraction peaks were composed of W and WS$_2$, showing that the solid lubricant WS$_2$ film was formed.

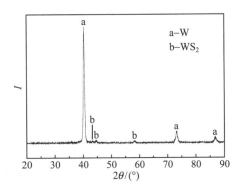

Fig.6.2   XRD patterns of the synthetic WS$_2$ film

Residual stresses are inevitably present in the film during deposition. In the sputtering process, the residual stresses are mainly generated as the result of ions bombardment and due to the difference of thermal and elastic properties between the film and substrate.

The residual stress was measured with X-ray stressometer using $\sin^2 \Psi$ method. Figure 6.3 shows the relation between 2θ and $\sin^2 \Psi$ for single W and synthetic WS$_2$ films, θ is the scanning angle, Ψ is side-tipping angle. The compressive residual stress measured on the W film was −48MPa, after sulfuration, the compressive stress on the WS$_2$ film decreased to −26MPa. The low residual stress can improve the bonding between the synthetic WS$_2$ film and the substrate.

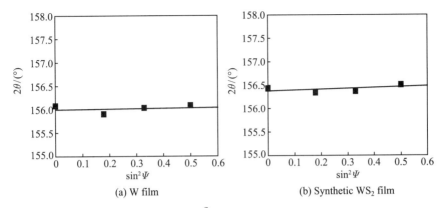

Fig.6.3   Relationships between 2θ and $\sin^2\Psi$

Figure 6.4 shows the load-displacement curves of the W and synthetic WS$_2$ film.

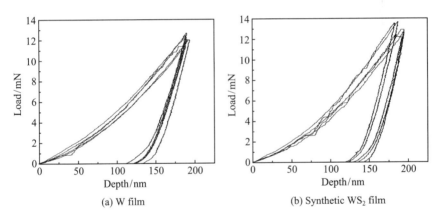

Fig.6.4   Load-displacement curves of the W and synthetic WS$_2$ film

Table 6.1 shows the nano-hardness and elastic modulus of the films and substrate.

Table 6.1   Nano-hardness and elastic modulus of the films and substrate

Sample	1045	FeS film	W film	WS$_2$ film
Nano-hardness/GPa	10.51	2.84	14.96	15.36
Elastic modulus/GPa	247.76	52.97	282.80	353.97

6.1.2   Tribological Properties of the Synthetic WS$_2$ Film

Friction and wear tests were conducted on a ball-on-disk wear tester under dry and oil lubrication conditions. The upper samples were 52100 steel balls, with a diameter of 6.35mm and hardness of HV 770. The lower samples were 1045 steel disks with WS$_2$ film, sulfurized and unsulfurized disks. Experimental parameters under dry condition were: load 5N, sliding speed 0.2m/s, and experimental time 60min. Experimental parameters under engine oil lubrication were: load 20N, sliding speed 0.2m/s, and experimental time 2h.

The friction force at high temperature was measured on a MMS-1G high temperature tester, whose largest silding speed and load are 100m/s and 450N, respectively. The experimental parameters were: temperature 600°C, sliding speed 5m/s, load 10N.

6.1.2.1   Tribological properties of the synthetic WS₂ film under dry condition

Figure 6.5 shows the tribological properties curves of the synthetic WS₂ film, FeS film, and 1045 steel. At the initial stage of the test, the friction coefficient of the synthetic WS₂ film was much lower than that of FeS film and 1045 steel, after 40min, the film was worn away, leading to the sudden rise and great fluctuation of friction coefficient. The wear scar depth of the WS₂ film increased with time, and it was obviously lower than that of the FeS film and 1045 steel in the whole test.

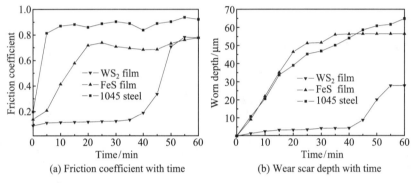

(a) Friction coefficient with time   (b) Wear scar depth with time

Fig.6.5 Tribological properties curves of the WS₂ film, FeS film, and 1045 steel under dry condition

Figure 6.6 shows the worn morphologies and EDX analysis results of 1045 steel, FeS film and synthetic WS₂ film sliding 60min under dry condition. Wide and deep worn scars could be seen clearly on the worn 1045 steel surface, the FeS film was entirely destroyed (almost no sulfur existed on the surface), but there were still a lot of WS₂ fragments remained on the worn surface of the synthetic WS₂ film (see Figure 6.6 (e) and (f)).

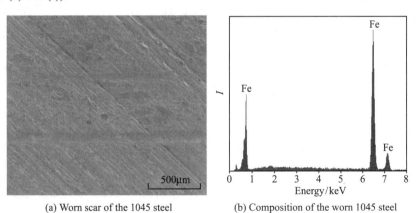

(a) Worn scar of the 1045 steel   (b) Composition of the worn 1045 steel

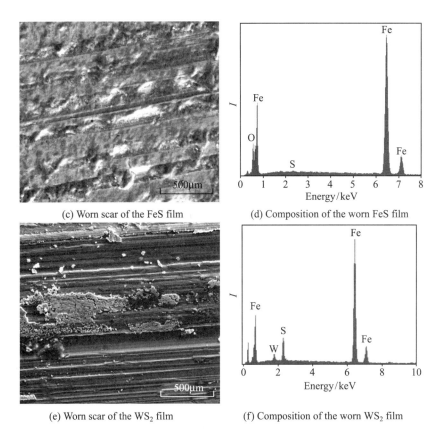

(c) Worn scar of the FeS film

(d) Composition of the worn FeS film

(e) Worn scar of the WS₂ film

(f) Composition of the worn WS₂ film

Fig.6.6 Worn scar and composition of the 1045 steel, FeS film and WS₂ film under dry condition

6.1.2.2 Tribological properties of the synthetic WS₂film under oil lubrication

Figure 6.7 shows the tribological properties curves of the WS₂ film, FeS film, and

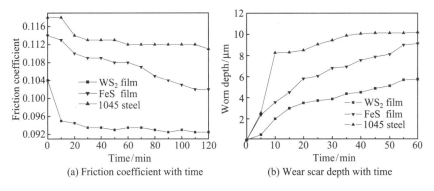

(a) Friction coefficient with time

(b) Wear scar depth with time

Fig.6.7 Tribological properties curves of the WS₂ film, FeS film, and 1045 steel under oil lubrication condition

1045 steel. During the whole test, the friction coefficient of the WS$_2$ film was stable and obviously lower than that of the FeS film and 1045 steel. The wear scar depth of the WS$_2$ film was always lower than that of other two samples.

Figure 6.8 shows the worn scars and EDX analysis results of 1045 steel, FeS film

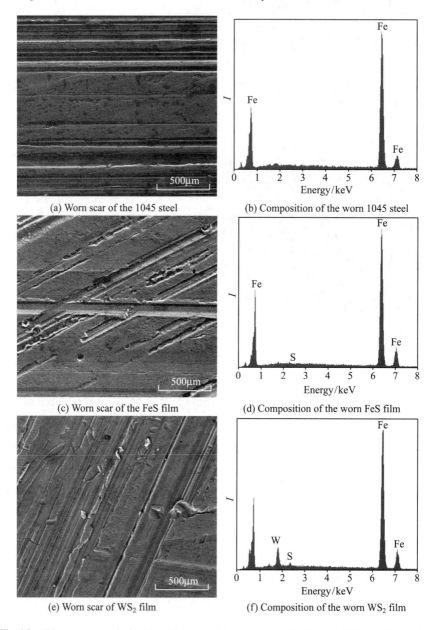

(a) Worn scar of the 1045 steel   (b) Composition of the worn 1045 steel

(c) Worn scar of the FeS film   (d) Composition of the worn FeS film

(e) Worn scar of WS$_2$ film   (f) Composition of the worn WS$_2$ film

Fig.6.8　Worn scar morphologies and compositions of the 1045 steel, FeS film and WS$_2$ film under oil lubrication condition

and synthetic WS$_2$ film sliding for 120min under oil lubrication condition. The furrows were generated on the surface of all three samples, but the wear of WS$_2$ film was less severe than that of FeS film and 1045 steel. The EDX analysis results demonstrated that no sulfur peaks existed on the worn FeS film surface, while certain sulfur could be detected on the worn surface of the synthetic WS$_2$ film.

6.1.2.3　Tribological properties of the synthetic WS$_2$ film at high temperature

The friction and wear tests were conducted on a high temperature and high speed testing machine of model MMS-1G.

Figure 6.9 shows the variation of friction coefficient of the synthetic WS$_2$ film, W film and M2 steel at high temperature of 600°C. The friction coefficient of the WS$_2$ varied in a very small range. It was stable and always less than that of the W film and M2 steel.

Figure 6.10 shows the wear loss of three samples. They were 0.5mg, 1.1mg and 1.7mg, respectively. It is obvious that the wear loss of the WS$_2$ film is much lower than that of the M2 steel and W film.

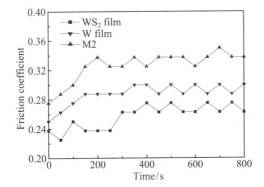

Fig.6.9　Variation of friction coefficient with time

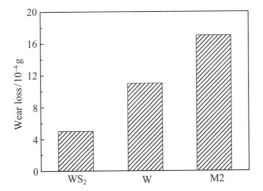

Fig.6.10　Wear loss of three samples

The worn morphologies of the M2 steel, W film and synthetic WS$_2$ film are shown in Figure 6.11. The worn surface of M2 steel was very rough and a large area of W film was flaking off, due to the absence of lubrication. While the worn surface of the WS$_2$ film was very smooth without any spalling symbol, meanwhile no obvious wear scar was present.

(a) M2 steel   (b) W film

(c) Synthetic WS$_2$ film

Fig.6.11   Worn morphologies of the M2 steel, W film and synthetic WS$_2$ film

6.1.2.4   Lubrication mechanism of the synthetic WS$_2$ film

Similar to the solid lubricant MoS$_2$ and graphite, WS$_2$ also possesses the close-packed hexagonal laminar structure, with one W atom linking two S atoms. In the friction process, due to the low shearing strength between the S atoms, the film can easily slip along the S—S planes. After a certain time of running-in, the transfer film can be generated on the surface of counterpart, therefore, the contact surfaces are both covered by the solid lubricant, which can prevent the direct contact between the metals and the occurrence of adhesion and scuffing. Meanwhile the preferred orientation of the surface film may be formed as shown in Figure 6.12 and the WS$_2$ film has a higher hardness and elastic modulus; in addition, WS$_2$ is very stable at high temperature, under the friction heat, it can be oxidized slowly to generate WO$_3$ protective layer, which possesses low friction coefficient and can prevent the further oxidization, all these factors are beneficial to improve the tribological properties of the WS$_2$ film.

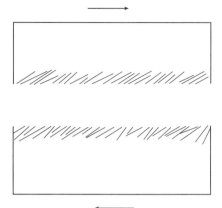

Fig.6.12   Preferred orientation of the surface film

6.2   WS₂/Ag Composite Film

As the unsaturated dangling bonds on the crystal edge of transition-metal sulfides (2H-MoS_2, 2H-WS_2, etc.) possess chemical activity, the sulfides are easily bonded to the metal surface and oxidized to decrease their tribological properties in humid air and oxygen-rich environments. When a small amount of soft metal Ag is added to WS_2 films, its wear-resistance can be improved.

6.2.1   Structures of WS₂/Ag Composite Film

Figure 6.13 shows the HRTEM morphology of the WS_2/Ag composite film. In the WS_2 film, Ag is present in the form of nanocrystal, while WS_2 shows poor crystallinity and basically exhibits amorphous structure [3].

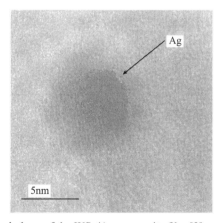

Fig.6.13   HRTEM morphology of the WS₂/Ag composite film [3]

Different Ag content has a big effect on the morphology of WS_2/Ag composite film [4]. The surface morphologies of WS_2/Ag composite film with different Ag contents are shown in Figure 6.14. The surface of pure WS_2 is loose, rough and porous; while the film becomes smooth and dense after the addition of Ag. The film surface is the densest when the Ag content reaches 5.05%.

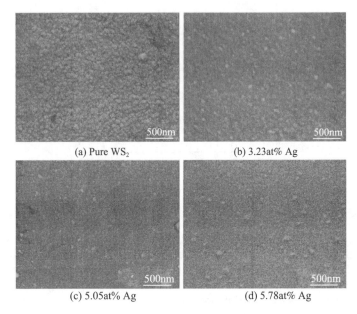

(a) Pure WS$_2$ (b) 3.23at% Ag

(c) 5.05at% Ag   (d) 5.78at% Ag

Fig.6.14 Surface SEM morphologies of WS$_2$/Ag composite films with different Ag content [4]

During deposition, different sputtering gas pressure also has great effect on the morphology and density of WS$_2$/Ag film, as shown in Figure 6.15 [3]. With the increase of sputtering gas pressure, the film density decreases but the film thickness is increased. When the pressures are 0.6Pa, 0.8Pa, 1.0Pa and 1.2Pa, the thicknesses of WS$_2$/Ag film are 0.71μm, 0.95μm, 1.22μm and 1.36μm, respectively. The white particles in Figure 6.15 (a) are Ag with the size of 80nm, which exhibit homogenous dispersion distribution in the WS$_2$ matrix. As the sputtered gas pressure becomes larger, the particle size increases; meanwhile, the film becomes loose (see Figure 6.15 (b)). This

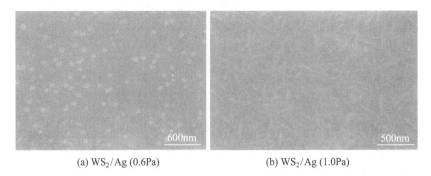

(a) WS$_2$/Ag (0.6Pa)   (b) WS$_2$/Ag (1.0Pa)

Fig.6.15 Surface SEM morphologies of WS$_2$/Ag films under different gas pressures [3]

is due to that the collision probability of the sputtered particles and discharged gas molecules is increased during the sputtering process, the scattering probability of particles becomes enhanced, so that the sputtering energy decreases, which leads to the decrease of bonding strength between the film and substrate, and to form a loose structure.

The bonding strength between the film and substrate is one of the most important indexes to evaluate the film quality [4]. Table 6.2 shows the impact of Ag content on the bonding strength between the film and substate. The bonding between the WS_2 film and substrate is relatively low, it can be obviously increased after the addition of Ag; When the Ag content is 5.05at%, it reaches the biggest bonding strength.

Table 6.2   Bonding strength between the film and substate at different Ag contents [4]

Ag content/at%	Critical load/N
0	20
3.23	28
5.05	36
5.78	31

The different content of Ag also has a certain effect on the hardness and elastic modulus of the WS_2/Ag composite film [5]. Figure 6.16 shows the variation of the hardness and elastic modulus with the Ag content for the WS_2/Ag composite film. With the addition of Ag in the film, the hardness and elastic modulus are prominently increased compared with those of the simple WS_2 film.

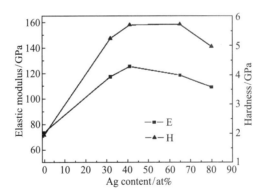

Fig.6.16   Variation of the hardness and elastic modulus with the Ag content for the WS_2/Ag composite film [5]

6.2.2   Tribological Properties of the WS₂/Ag Composite Film

Figures 6.17 and 6.18 are the variation curves of friction coefficients for the simple WS_2 film and WS_2/Ag composite film with different Ag contents in atmosphere and vacuum [4]. The friction coefficient of the simple WS_2 film was large and unstable; meanwhile, the fluctuation was relatively severe, while the friction coefficient of WS_2/Ag composite film was lower than that of simple WS_2 film. The composite film with 5.05at% Ag showed the smallest friction coefficient in atmosphere and vacuum.

The tribological curves of all films in vacuum were smoother and lower than those in atmosphere, this is due to that the films are easily subject to the effect of water and oxygen in atmosphere during friction, and partial oxidization and deliquescence of films may be present.

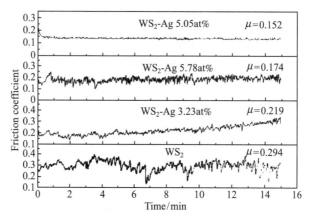

Fig.6.17 Friction coefficients of films in atmosphere [4]

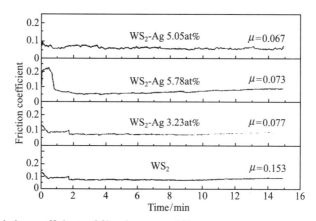

Fig.6.18 Friction coefficients of films in vacuum [4]

Figure 6.19 shows the variation of friction coefficient of the composite film with number of cycles in humid air [6]. The fluctuation of the friction coefficient of WS$_2$/Ag composite film is lower than that of simple WS$_2$ film during the friction and wear process, indicating that the appropriate addition of Ag to the films can result in a decrease of friction coefficient and improvement of wear stability.

Figure 6.20 shows the variation of friction coefficients with time for the simple WS$_2$ and WS$_2$/Ag composite films in vacuum and ambient atmosphere [3]. The friction coefficient of WS$_2$/Ag composite film deposited under the same condition was slightly higher than that of simple WS$_2$ film. The friction coefficient of all films in vacuum was lower than that in ambient atmosphere. Although the friction coefficient of WS$_2$ film was lower, its fluctuation was greater, indicating that the friction state of WS$_2$/Ag composite film was more stable.

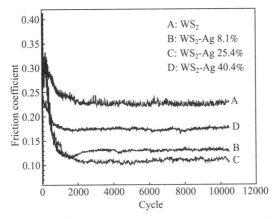

Fig.6.19   Variation curves of friction coefficient of the composite films with number of cycles, tested in humid air [6]

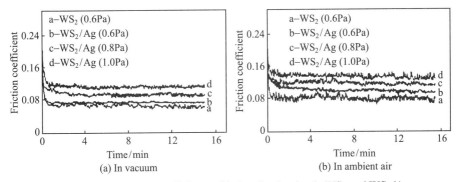

Fig.6.20   Variation of friction coefficients with time for the simple WS$_2$ and WS$_2$/Ag composite films [3]

The sputtering gas pressure is also one of the factors affecting the friction coefficient of the WS$_2$/Ag composite film [4]. The friction coefficients of WS$_2$/Ag composite films prepared under different sputtering gas pressures are shown in Figure 6.21. The friction coefficients went up with the increase of sputtering gas pressure, the lower the pressure, the denser the film; the higher the bonding strength and the better the tribological properties.

Figure 6.22 shows the effect of Ag content on wear resistance of the composite film [6]. The addition of Ag into the film enhances the wear resistance of the film obviously, and the WS$_2$/Ag film with 16.2% Ag has the highest wear resistance.

Figure 6.23 shows the worn SEM morphologies of WS$_2$/Ag and WS$_2$ films tested in the humid atmosphere [3]. The wear scar of the WS$_2$ film was deep, and some cracks were present, while the worn surface of the WS$_2$/Ag film was smoother with tiny wear scar, and no obvious cracks occurred. Besides the effect of bonding strength and density, this is also related with the Ag phase dispersed in the WS$_2$ film. In the friction and wear process, the Ag-rich surface layer was formed due to the plastic flow of Ag on the film surface, which can prevent the shearing deformation of WS$_2$ film and improve the resistance of crack initiation and propagation. As a result, the wear life can be extended.

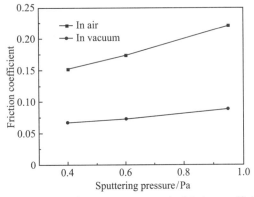

Fig.6.21   Effect of different sputtering gas pressure on the friction coefficient of WS$_2$/Ag composite film [4]

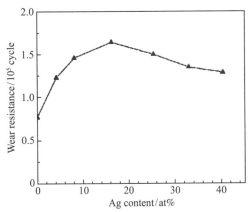

Fig.6.22   Dependence between wear resistance and Ag content of the composite films [6]

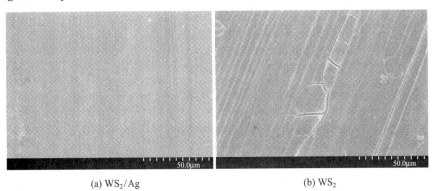

(a) WS$_2$/Ag   (b) WS$_2$

Fig.6.23   Worn SEM morphologies of WS$_2$/Ag film and WS$_2$ film in humid atmosphere [3]

6.3   WS$_2$/MoS$_2$ Multilayer Film

6.3.1   WS$_2$/MoS$_2$ Co-sputtered Film

Many researchers have deeply studied WS$_2$/MoS$_2$ co-sputtered film. The results showed that the tribological properties of WS$_2$/MoS$_2$ co-sputtered film were superior to those of single films. Therefore, the WS$_2$/MoS$_2$ co-sputtered films have a good application prospect.

Figure 6.24 shows a TEM image of the WS$_2$/MoS$_2$ multilayer film, clearly demonstrating that the film has been grown as a continuous layered structure [7].

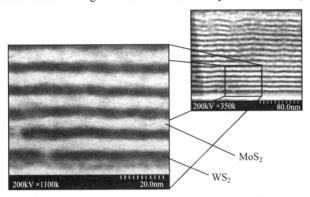

Fig.6.24   Cross-sectional TEM image of the WS$_2$/MoS$_2$ multilayer film [7]

Figure 6.25 shows the variation of friction coefficients with sliding distance for the WS$_2$/MoS$_2$ multilayer film and WS$_2$, MoS$_2$ single films in humid air and vacuum [8]. The friction coefficients of both the single-layer film and multilayer film are

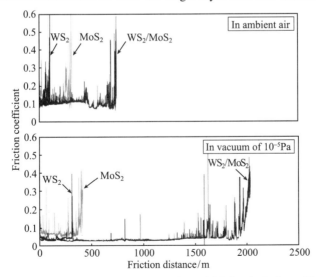

Fig.6.25   Variation of friction coefficients for WS$_2$ and MoS$_2$ single-layer films, and WS$_2$/MoS$_2$ multilayer film with the number of revolutions at room temperature in humid air and vacuum [8]

approximately 0.1 in the humid air, whereas they are less than half of that value (<0.05) in vacuum. In particular, the friction coefficient of the multilayer film in vacuum is close to 0.03, and its wear life is increased more than two times compared with that in the humid air. However, the two single-layer films do not show an obvious increase of wear life. Therefore, the friction-reducing and wear-resistance properties of the multilayer film are superior to those of the single-layer films both in vacuum and humid air.

Figure 6.26 shows the wear rate of the single-layer films and multilayer film [7]. The wear rate of the WS₂ single-layer film was equivalent to that of MoS₂, while the wear rate of the multilayer film was obviously lower than that of WS₂ and MoS₂.

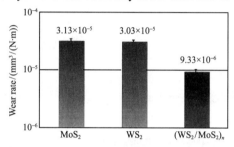

Fig.6.26   Wear rate of the single-layer films and multilayer film [7]

The tribological properties of co-sputtered Ni/WS₂/MoS₂ composite film are also superior to those of single-layer film, because the addition of Ni is helpful to the film growth parallel to the substrate surface [9]. Figure 6.27 shows the XRD patterns of MoS₂ single-layer film and Ni/WS₂/MoS₂ composite film. In the composite film, only the peak of (002) was present; while in the MoS₂ film, the peaks of (100) and (110) appeared but the peak of (002) was absent. It is indicated that the crystal planes of the two films were completely reverse: The basal plane (002) of the composite film was parallel to the substrate surface; while that of the MoS₂ film was perpendicular to the substrate surface. The tribological properties of the former were superior to those of the latter, because the former was more conducive to the friction-reduction and protection of the film from oxidation.

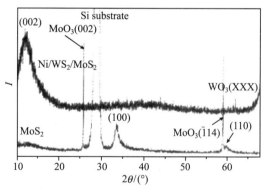

Fig.6.27   XRD patterns of MoS₂ single-layer film and Ni/WS₂/MoS₂ composite film [9]

Figure 6.28 shows the cross-section HR-TEM morphologies of single MoS₂ film and Ni/WS₂/MoS₂ composite film. It can be found that the structure of the composite film is much denser than that of the single MoS$_2$ film. The loose structure is easy to be oxidized and absorb moisture, thus the friction coefficient of the film will be rapidly increased causing the failure of lubrication.

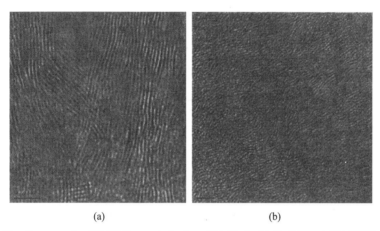

(a)   (b)

Fig.6.28   Cross-section HR-TEM morphologies of (a) Single MoS$_2$ film and (b) Ni/WS$_2$/MoS$_2$ composite film [9]

Figure 6.29 shows the variation of friction coefficient of MoS$_2$ film and Ni/MoS$_2$/WS$_2$ composite film with time at 25N load. The failure of the MoS$_2$ film happened in a very short time, while the wear life of the composite film achieved 42000 revolutions, about 40 times longer than that of the former.

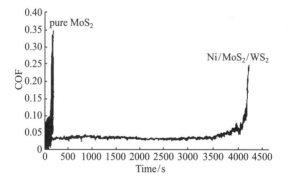

Fig.6.29   Variation of friction coefficient of MoS$_2$ film and Ni/MoS$_2$/WS$_2$ composite film at 25N load [9]

6.3.2   WS$_2$/MoS$_2$ Multilayer Film Prepared by Combined Treatment

The substrate materials were 1045 steel and M2 steel. The specimens used for T-11 ball-on-disc tester were 1045 discs with dimensions of ϕ25.4mm×6mm and hardness of HRC 55; The specimens used for MMS-1G high temperature tester were cylindrical pins with dimensions of ϕ14mm×40.3mm and hardness of HRC 63—64. Two W/Mo multilayer films with thickness of 1.8μm and 1.0μm were prepared on the 1045 steel and M2 steel by magnetron sputtering + low temperature ion sulfuration complex treatment.

The W and Mo targets of 99.99% purity were made with the dimensions of 124mm×254mm×2mm and 124mm×254mm×10mm, respectively. The six layers W/Mo multilayer films were deposited alternatively by W and Mo with 300nm thickness for each layer and 1.8μm thickness for total film. The two W/Mo multilayer films were deposited with 500nm thickness for each layer and 1.0μm thickness for total film. The two W/Mo multilayer films were then treated by ion sulfuration for 2h.

6.3.2.1   Characterizations of the WS$_2$/MoS$_2$ multilayer film

Figure 6.30 shows the surface morphology of the WS$_2$/MoS$_2$ multilayer film. The surface of the film was quite smooth, with plenty of particles. Some of them were piled up to form agglomerated particles; the largest dimension of them could reach 5μm.

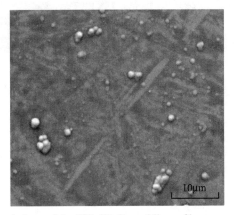

Fig.6.30   Surface morphology of the WS$_2$/MoS$_2$ multilayer film

Figure 6.31 shows the XRD pattern of the WS$_2$/MoS$_2$ multilayer film. The diffraction peaks are mainly composed of W, Mo, WS$_2$ and MoS$_2$. W/Mo multilayer film possesses plenty of grain and layer boundaries, which are beneficial to the diffusion of sulfur into W and Mo layers.

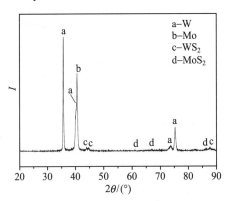

Fig.6.31   XRD pattern of the WS$_2$/MoS$_2$ multilayer film

Table 6.3 shows the data of nano-hardness and elastic modulus for films.

Table 6.3   Nano-hardness and elastic modulus of films

Film	W film	Mo film	WS$_2$ film	MoS$_2$film	WS$_2$/MoS$_2$ multilayer film
Nano-hardness/GPa	15.22	2.90	15.36	1.50	14.84
Elastic modulus/GPa	176.64	223.45	353.97	162.20	263.87

6.3.2.2   Tribological properties of the WS$_2$/MoS$_2$ multilayer film

1. Tribological properties of the WS$_2$/MoS$_2$ multilayer film under dry condition

Figure 6.32 shows the tribological property curves of the WS$_2$/MoS$_2$ multilayer film, WS$_2$ film, and 1045 steel under dry condition. The friction coefficient of the WS$_2$/MoS$_2$ multilayer film was quite low and stable during the whole test, and it was obviously lower than that of the WS$_2$ film and 1045 steel. The wear scar depth of the WS$_2$/MoS$_2$ multilayer film was much less than that of WS$_2$ film and 1045 steel as well.

Figure 6.33 are the worn morphologies and compositions of the WS$_2$/MoS$_2$ multilayer film after 5min and 60min sliding respectively under dry condition. The worn surface of the WS$_2$/MoS$_2$ multilayer film was very smooth and featureless after 5min

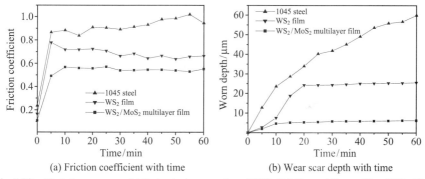

(a) Friction coefficient with time   (b) Wear scar depth with time

Fig.6.32   Variation curves of tribological properties of WS$_2$/MoS$_2$ multilayer film, WS$_2$ film, and 1045 steel under dry condition

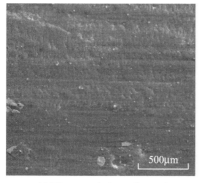
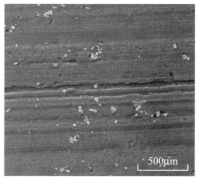

(a) Worn morphology after 5min   (b) Worn morphology after 60min

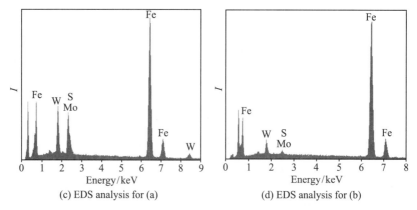

Fig.6.33 Worn scar morphologies and EDS analysis of the WS₂/MoS₂ multilayer film at different moment under dry condition

sliding. A small amount of furrows were present after 60min sliding, the film had basically been worn out, but some film patches still remained on the worn surface.

2. Tribological properties of the WS₂/MoS₂ multilayer film under oil lubrication condition

Figure 6.34 shows the tribological property curves of the WS₂/MoS₂ multilayer film, WS₂ film, and 1045 steel under oil lubrication condition. The friction coefficient of WS₂/MoS₂ multilayer film was much lower than that of WS₂ film and 1045 steel. It varied in the range of 0.085—0.1. The wear scar depth of multilayer film was only 3μm, much lower than that of WS₂ film (5μm) and 1045 steel (10μm).

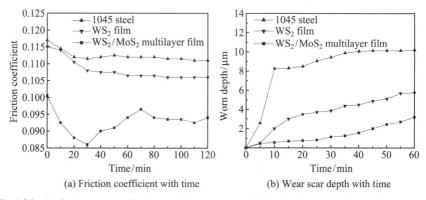

Fig.6.34 Variation curves of tribological properties of WS₂/MoS₂ multilayer film, WS₂ film, and 1045 steel under oil lubrication condition

Figure 6.35 shows the worn morphologies and compositions of the 1045 steel, WS₂ film and WS₂/MoS₂ multilayer film after 120min test under oil lubrication condition. A lot of furrows were present on the worn 1045 steel surface; the wear scar depth of the WS₂ film was relatively shallow. The worn surface of the WS₂/MoS₂ multilayer film was smooth; meanwhile, a certain amount of sulfur could be detected.

3. Tribological properties of the WS₂/MoS₂ multilayer film under high temperature

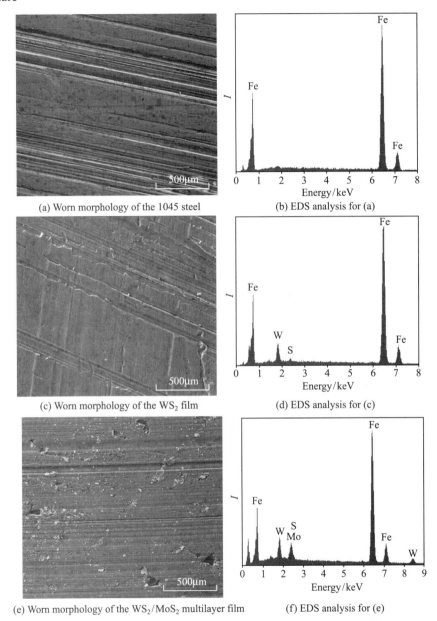

(a) Worn morphology of the 1045 steel   (b) EDS analysis for (a)

(c) Worn morphology of the WS₂ film   (d) EDS analysis for (c)

(e) Worn morphology of the WS₂/MoS₂ multilayer film   (f) EDS analysis for (e)

Fig.6.35   Worn scar morphologies and EDS analysis under oil lubrication condition

Figure 6.36 shows the variation of friction coefficient with time for the M2 steel, W/Mo film and WS₂/MoS₂ multilayer film under high temperature of 600°C. At the beginning, the friction coefficients of the three samples had no big difference, but when the test ended, the friction coefficient of the multilayer film was much lower than that of other two samples, indicating that the lubrication role of the multilayer film became more and more obvious. During the whole test, the friction coefficient of the multilayer film was always the lowest among three samples.

Figure 6.37 shows the wear loss of three samples. The wear loss of them were 1.7mg, 1.4mg and 0.6mg, respectively. The wear loss of the M2 steel was nearly three times that of the multilayer film, while the wear loss of the W/Mo film was more than two times that of the multilayer film.

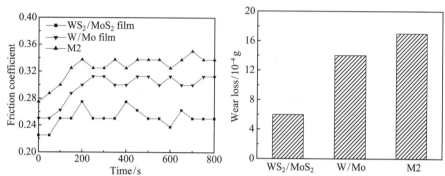

Fig.6.36 Variation of friction coefficient with time

Fig.6.37 Wear loss of three samples

Figure 6.38 shows the worn morphologies of the M2 steel, W/Mo film and WS₂/MoS₂ multilayer film. Obvious oxidation and plastic deformation occurred on the

(a) M2 steel   (b) W/Mo film

(c) WS₂/MoS₂ multilayer film

Fig.6.38　Worn morphologies of the M2 steel, W/Mo film and WS₂/MoS₂ multilayer film

worn M2 steel; the wear was very severe. The wear of the W/Mo film was mild compared with that of the M2 steel; but partial film had flaked off; no obvious oxidation happened. The worn surface of the multilayer film was smooth, and no adhesion occurred.

6.3.2.3 Lubrication mechanism of the WS_2/MoS_2 multilayer film

Each layer in the multilayer film has its respective function. The MoS_2 film is softer; its friction-reduction property is very good, but it has poorer wear-resistance, while the WS_2 film is harder, its wear-resistance is better. The alternating structure of WS_2 and MoS_2 layers shows an effective synergistic role, i.e., it can overcome one's weakness by acquiring other's strong point.

In addition, the WS_2/MoS_2 multilayer film made by sputtering and sulfurizing combined treatment has a dense structure, which can effectively prevent the oxidation of the film.

6.4 WS_2/CaF_2 Composite Coating

Ni-based submicron WS_2/CaF_2 self-lubricating composite coatings were produced on carbon steel substrate by high velocity oxygen fuel (HVOF) spray processing [10]. The friction and wear behaviors of the coatings were carried out using a pin-on-ring tribometer (MMS-1G) at room temperature and 400°C, respectively.

Figure 6.39 shows the structures of the WS_2/CaF_2 self-lubricating composite coatings with a thickness of 400—450μm (Figure 6.39 (a)). The self-lubricating composite coating exhibits a less pores and lamellar structure containing melted, partially melted and a few unmelted powder particles (see Figure 6.39 (b)). The main structure defects in coating are micropores and microcracks, which are randomly distributed in the whole coating, and come fine chromium carbides can be found in some melted particles (see Figure 6.39 (c)). Some pores are present at the interface (see Figure 6.39 (d)), which may contribute to the fact that the partially melted particles do not fully fill the gaps of the rough substrate surface, or some particles do not bond to their neighboring particles.

Figure 6.40 (a) shows the friction coefficients of as-sprayed coating sliding against AISI 52100 bearing steel at ambient temperature. The friction coefficient of Ni45 coating is about 0.6. But the composite coatings exhibit a lower friction coefficient

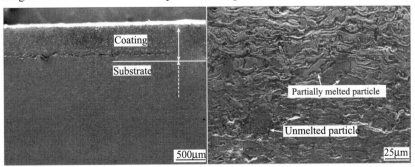

(a) As-sprayed coating   (b) Unmelted particles

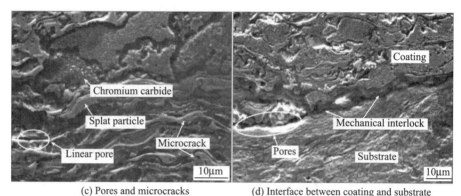

(c) Pores and microcracks   (d) Interface between coating and substrate

Fig.6.39   Typical morphologies of as-sprayed self-lubricating composite coating [10]

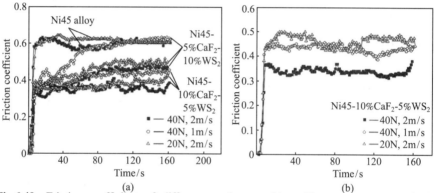

(a)   (b)

Fig.6.40   Friction coefficient of different coatings at (a) ambient temperature and (b) 400°C [10]

with comparison to the coating without solid lubricants. For Ni45-5%CaF₂-10%WS₂ and Ni45-10%CaF₂-5%WS₂ coatings, the friction coefficients at room temperature are in the range of 0.35—0.48 and 0.31—0.41, respectively, which are carried out under a load of 40N and a speed of 2m/s. When the testing velocity decreases to 1m/s, the friction coefficients of both Ni45-5%CaF₂-10%WS₂ and Ni45-10%CaF₂-5%WS₂ coatings increase to 0.40—0.63 and 0.35—0.46, respectively. This indicates that the testing velocity has significant influence on tribological properties of coating. A lower friction velocity may lead coatings not to transfer solid lubrication film onto interfaces.

The friction coefficient of Ni45-10%CaF₂-5%WS₂ coating at 400°C indicates that temperature has a little influence on the frictional properties of the coatings, as shown in Figure 6.40 (b). Under a load of 40N and a speed of 2m/s, the friction coefficient fluctuates in 0.32—0.38, and the friction coefficient of the other operating conditions is higher than that at room temperature.

6.5   Ni-P-(IF-WS₂) Composite Film

For Ni-P and Ni-P-(IF-WS₂) composite films, after annealing at 673K, Ni₃P hard phase can precipitate from the matrix; which promotes the transformation from amorphous to crystalline state, so that the hardness and performance of the film can be improved [11]. Figure 6.41 shows the tribological properties of Ni-P and Ni-P-(IF-WS$_2$) composite films after annealing in atmosphere. The wear rates of the two films are increased with load, but the wear rate of the composite film is much less than that of the Ni-P coating. When the load becomes larger, the friction coefficient of the Ni-P film is increased, while that of the composite film decreases slowly.

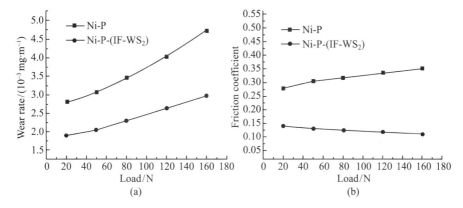

Fig.6.41 Tribological properties of Ni-P and Ni-P-(IF-WS$_2$) composite films after annealing [11]

Figure 6.42 shows the worn morphologies of Ni-P and Ni-P-(IF-WS$_2$) composite films [12]. For the Ni-P-(IF-WS$_2$) composite film, only slight wear trace can be seen on the worn surface; while for the Ni-P coating, obvious wear trace is present, and a little brittle debris peeled off from the film surface. The mechanism of that IF-WS$_2$ nano material can significantly improve the tribological properties can be explained as follows: In the friction process, IF-WS$_2$ nano-particles in the composite film are released from the bulk phase to the metal surface, they can improve the contact between friction surfaces as the third body; meanwhile, the nano material with spherical or similar to spherical particles can roll and slide between friction surfaces freely, so that the wear rate and friction coefficient of the material can be decreased effectively.

(a) Ni-P-(IF-WS$_2$) composite film　　　　　(b) Ni-P film

Fig.6.42　Worn morphologies of the two films [12]

　　Table 6.4 shows the test results of tribological properties for Ni-P, Ni-P-(2H-WS$_2$), Ni-P-(IF-WS$_2$) and Ni-P-graphite films after heat treatment in vacuum for 2h [12]. Ni-P-(IF-WS$_2$) composite film possesses lower friction coefficient compared with the other films; it also has excellent wear-resistance. Such composite film can not only improve the service life of friction-pairs, but also effectively decrease the energy consumption of machines. Therefore, it has an extensive application prospect in modern industries and aeronautical and space technologies.

Table 6.4　Test results of tribological properties for films [12]

Coating	Mass loss of block/mg	Friction coefficient
Ni-P	15.6	0.090
Ni-P-(2H-WS$_2$)	5.2	0.062
Ni-P-graphite	4.3	0.067
Ni-P-(IF-WS$_2$)	3.0	0.030

References

1. Zhu L N, Li G L, Wang H D, et al. Microstructures and nano mechanical properties of the metal tungsten film [J]. Current Applied Physics, 2009, 9: 510–514.
2. Zhu L N, Wang C B, Wang H D, et al. Tribological properties of WS$_2$ composite film prepare by a two-step method [J]. Vacuum, 2010, 85: 16–21.
3. Lai D M, Xu J P, Zhang S C, et al. Friction and wear properties of sputtered WS$_2$/Ag nanocomposite films in different environments [J]. Tribology, 2006, 26(6): 515–518.
4. Peng S M. Tribological behaviours of WS$_2$/Ag films and Mo sulfide/Ti films [D]. Hangzhou: Zhejiang University, 2007.
5. Wang Q, Tu J P, Zhang S C, et al. Effect of Ag content on microstructure and tribological performance of WS$_2$–Ag composite films [J]. Surface & Coatings Technology, 2006, 201: 1666–1670.
6. Zheng X H, Tu J P, Lai D M, et al. Microstructure and tribological behavior of WS$_2$-Ag composite films deposited by RF magnetron sputtering [J]. Thin Solid Films, 2008, 516: 5404–5408.
7. Watanabe S, Noshiro J, Miyake S. Tribological characteristics of WS$_2$/MoS$_2$ solid lubricating multilayer films [J]. Surface and Coatings Technology, 2004, 183: 347–351.
8. Watanabe S, Noshiro J, Miyake S. Friction properties of WS$_2$/MoS$_2$ multilayer films under vacuum environment [J]. Surface & Coatings Technology, 2004, 188–189: 644–648.
9. Yin G L, Huang P H, Yu Z, et al. Microstructure and tribological performance of co-sputtered Ni/MoS$_2$/WS$_2$ composite films [J]. Journal of Harbin Institute of Technology, 2006, 38: 151–153.
10. Zhang X F, Zhang X L, Wang A H, et al. Microstructure and properties of HVOF sprayed Ni-based submicron WS$_2$/CaF$_2$ self-lubricating composite coating [J]. Transactions of Nonferrous Metals Society of China, 2009, 19: 85–92.
11. Zou T Z, Xia Z Z, Tu J P, et al. Preparation and friction and wear properties of Ni-P-(IF-WS$_2$) electroless composite coating [J]. Materials Protection, 2004, 37(7): 91–93.
12. Chen W X, Tu J P, Ma X C, et al. Preparation and tribological properties of Ni-P electroless composite coating containing inorganic fullerene-like WS$_2$ nanomaterials [J]. Acta Chimica Sinica, 2002, 60(9): 1722–1726.

Chapter 7

Micron-nano ZnS Solid Lubrication Film

ZnS possesses close-packed hexagonal lattice structure with low shearing strength; it is also an effective solid lubricant. At present, a lot of studies have been done on ZnS nano-particles used as the additive in lubricating oil. However, still few people have conducted the researches on the ZnS film used as a solid lubricant.

7.1 ZnS Film Prepared by High Velocity Arc Spraying + Sulfurizing Treatment

The soft metals used commonly as the solid lubricant mainly include Pb, Zn, Sn, In, Au, and Ag, etc.; Zn has been applied widely because of its low cost. The electroplating or hot-dip methods are usually used to prepare the Zn and other soft metal layers.

Here the high velocity arc spraying technique was utilized to prepare Zn layer on the 1045 steel; then it was treated by low temperature ion sulfurizing to form ZnS/Zn composite layer. Under dry condition, the friction-reduction and wear-resistance properties were improved due to the combined effect of Zn and ZnS, but the lubrication performance was very poor under oil lubrication condition.

7.1.1 Preparation

The substrate material was 1045 steel with the hardness of HRC 55. The surface roughness was R_a=0.8μm. Before deposition, the substrate was pretreated by grit-blasting; then the Zn layer was sprayed using an HAS-01 spraying gun. The spraying parameters were: spraying voltage 35V, spraying current 160A, pressure of compressed air 0.7MPa, and spraying distance 150mm. A 300μm thick Zn layer was prepared finally.

The Zn layer was then treated by ion sulfurizing for 2h.

7.1.2 Characterizations

7.1.2.1 Morphologies

Figure 7.1 shows the surface, cross-section morphologies and distribution of elements. The surface of the Zn/ZnS composite layer was relatively compact and smooth with a few particles. The thickness of ZnS layer was about 3μm.

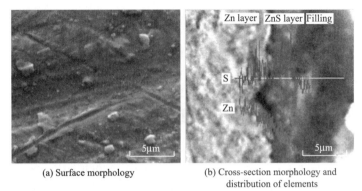

(a) Surface morphology

(b) Cross-section morphology and
distribution of elements

Fig.7.1   Surface, cross-section morphologies and distribution of elements

7.1.2.2   Phase structure

Figure 7.2 shows the XRD pattern of the Zn/ZnS composite layer. The diffraction peaks were mainly composed of ZnS and Zn.

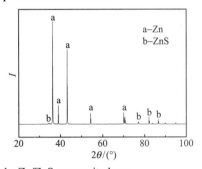

Fig.7.2   XRD pattern of the Zn/ZnS composite layer

7.1.2.3   Nano-mechanical properties

Figure 7.3 shows the load-displacement curves of the Zn layer and Zn/ZnS composite layer.

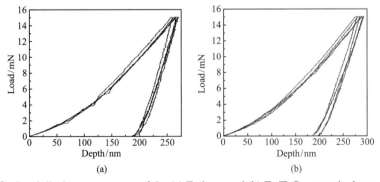

Fig.7.3   Load-displacement curves of the (a) Zn layer and (b) Zn/ZnS composite layer

The nano-mechanical properties of all the specimens are given in Table 7.1.

Table 7.1   Nanohardness and elastic modulus of all specimens

Specimen	1045 steel	FeS layer	Zn coating	Zn/ZnS composite layer
Nano-hardness/GPa	10.51	2.84	9.35	3.86
Elastic modulus/GPa	247.76	52.97	179.49	123.25

7.1.3   Tribological Properties

The friction and wear tests were performed on a T-11 ball-on-disk tester under dry and oil lubrication conditions. The upper sample was 52100 steel ball; the lower sample was the 1045 steel disc with composite Zn/ZnS layer. The sulfurized 1045 steel and original steel discs were also tested under the same conditions for comparison. Experimental parameters under dry friction condition were: load 5N, sliding speed 0.2m/s, and experiment time 60min; experimental parameters under oil lubrication were: loads 20N, 30N and 40N, respectively, sliding speed 0.2m/s, and experimental time 2h.

7.1.3.1   Tribological performances under dry condition

Figure 7.4 shows the tribological curves of the 1045 steel, FeS film, and Zn/ZnS composite layer under dry condition. The friction coefficient of the Zn/ZnS layer was extremely low and stable, ranging from 0.1 to 0.15; in the whole test, it was much lower than that of the 1045 steel and FeS film. The wear scar depth of the Zn/ZnS layer slowly increased and achieved a stable state finally; the wear scar depth of the 1045 steel increased linearly; the wear scar depth of the FeS film increased fast at first, and then tended to be stable, but it was still much larger than that of the Zn/ZnS layer.

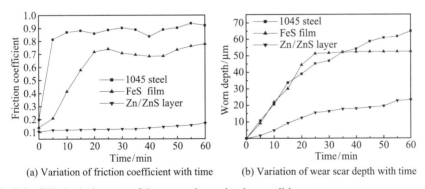

(a) Variation of friction coefficient with time   (b) Variation of wear scar depth with time

Fig.7.4   Tribological curves of three samples under dry condition

Figure 7.5 shows the worn morphologies and compositions of the 1045 steel, FeS film and Zn/ZnS composite layer after sliding of 60min under dry condition. The wear of the 1045 steel was severe; obvious wear scar was present. The wear of the FeS film was milder compared with that of the 1045 steel, but the FeS film of large area had flaked off. The wear scar of the Zn/ZnS layer was the slightest; only a little part of the layer was worn out. A certain amount of sulfur still remained on the surface of Zn/ZnS layer at the end of test.

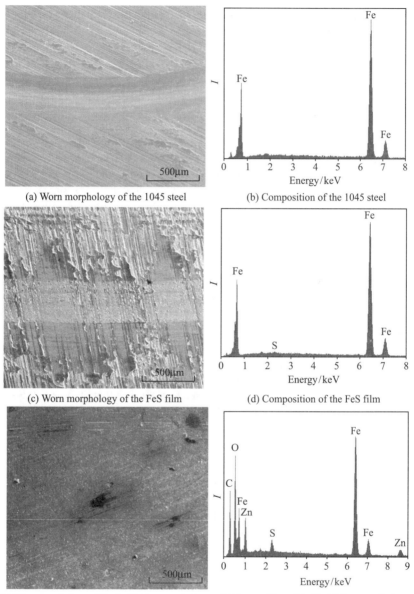

(a) Worn morphology of the 1045 steel   (b) Composition of the 1045 steel

(c) Worn morphology of the FeS film   (d) Composition of the FeS film

(e) Worn morphology of the Zn/ZnS composite layer (f) Composition of the Zn/ZnS composite layer

Fig.7.5 Worn morphologies and compositions of the 1045 steel, FeS film and Zn/ZnS composite layer under dry condition

7.1.3.2 Tribological performances under oil lubrication condition

The tribological properties of the Zn/ZnS composite layer were very poor, the friction coefficient and wear scar depth were significantly higher than those of the 1045 steel. The worn morphologies of the 1045 steel and Zn/ZnS layer under the load of 40N after sliding of 60min are shown in Figure 7.6. The wear scar of the Zn/ZnS layer was much deeper and wider compared with that of the 1045 steel; it seemed quite incredible that the tribological properties of the Zn/ZnS layer were inferior to those of the original steel.

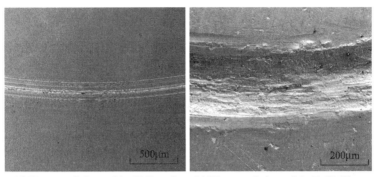

(a) Worn morphology of the 1045 steel   (b) Worn morphology of the Zn/ZnS layer

Fig.7.6   Worn morphologies of the 1045 steel and Zn/ZnS composite layer under oil lubrication condition

7.1.4   Lubrication Mechanisms of the Zn/ZnS Composite Layer

The Zn/ZnS composite layer possessed excellent tribological properties under dry condition; the main reasons were as follows:

(1) During the friction process, the Zn/ZnS composite layer is easily crushed and adhered to the surface of counterpart, which can effectively prevent the direct contact of metals and reduce the adhesion wear.

(2) When the ZnS layer is worn off, the Zn layer can continue to play the role of lubrication. In addition, the sprayed Zn layer possesses a porous structure, its pores can accommodate the wear debris making the contact surface clean.

However, the Zn/ZnS composite layer showed poor wear resistance under oil lubrication. This was possibly due to that the sprayed Zn layer had plenty of pores, which were mutually unconnected. When the layer was subject to the normal load, the oil stored in the pores could not overflow; meanwhile, the lubricating oil was incompressible. Therefore, the high oil pressure promoted the spalling of the ZnS and Zn layers.

7.2   ZnS Film Prepared by Nano-brush Plating + Sulfurizing Treatment

When aircraft is operating in space, it will encounter extremely harsh service conditions such as sharp alternating of high and low temperature, ultra-high vacuum, ultra-high load and ray radiation (ultraviolet, atomic oxygen), which makes the fluid lubricant been unable to effectively play a role in reducing friction. In these conditions, serious friction and wear due to poor lubrication becomes one of the main causes leading to the damage and failure of key components of spacecraft.

Here we explored nano-brush plating and low temperature sulfurizing combined techniques to obtain Zn/ZnS composite coating which is suitable for the application in the space conditions, and investigates the tribological properties and evolution mechanism in the space conditions (high vacuum, ultraviolet radiation, and atomic oxygen radiation).

7.2.1   Preparation

The substrate material was 1045 steel, and its hardness was HRC 55 after quenched. The surface roughness R_a was 0.8μm after polishing. Nano electro-brush plating was used to prepare Zn coating on the surface of quenched 1045 steel. Preparation process is as following: ① pretreatment; ② current purification; ③ strong activation; ④ weak activation; ⑤ plating base coat; ⑥ plating alkalinous Zn with a thickness of 100μm. Then the obtained Zn coating was treated by low temperature ion sulfuration for 2h (190°C). Finally, the ZnS film with a thickness of 2μm was fabricated. A composite Zn/ZnS layer with the base coat Zn and top coat ZnS was obtained.

Vacuum friction and wear tester was applied to investigate the tribological properties of composite Zn/ZnS layer under dry friction condition in four environments such as air, vacuum, atomic oxygen radiation and ultraviolet vacuum radiation. The upper samples were 9Cr18 steel balls, 3mm in diameter and with a hardness of HV 770. The lower samples were AISI 1045 steel discs with compound Zn/ZnS layers, 45mm in diameter and 8mm thick. During friction, the upper sample was fixed, and the lower disc rotated. The test conditions were as following: speed of rotation was 100r/min; load was 3N; friction time was 600s; degree of vacuum was 3.6×10^{-5}Pa; time of ultraviolet radiation was 10h; photon energy was higher than 376.6kJ/mol.

7.2.2   Morphologies

Figure 7.7 shows SEM images of the composite Zn/ZnS layer in air, vacuum, atomic oxygen radiation and ultraviolet radiation environments. From Figure 7.7 (a) and (b), it can be seen that the compound Zn/ZnS layer was compact and uniform in air and

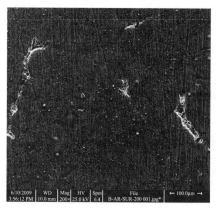
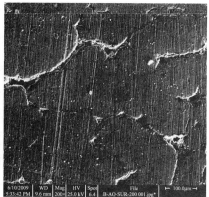

(a) In air   (b) In vacuum

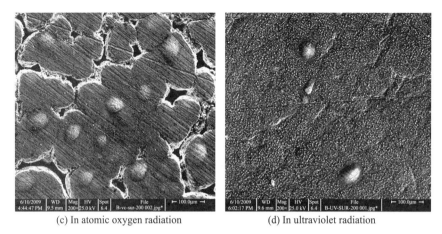

(c) In atomic oxygen radiation   (d) In ultraviolet radiation

Fig.7.7   SEM images of the composite Zn/ZnS layer in air, vacuum, atomic oxygen radiation and ultraviolet radiation environments

vacuum.   There were tiny white particles, and crystal cell boundaries could be observed clearly especially in vacuum environment. From Figure 7.7 (c) and (d), the morphologies were very different compared with the former two images. After atomic oxygen radiation for 5h, many big spherical particles with even distribution appeared on the layer surface. Lots of pores at grain boundaries were observed. After ultraviolet radiation, many tiny white particles were precipitated from the compound layer, and there were a few big particles, 20—50μm in diameter. In this environment, the compound layers were relatively compact, and no obvious crystal cell boundaries were found.

7.2.3   Friction Coefficient

The variation of friction coefficient with time for the compound Zn/ZnS layer in four environments is shown in Figure 7.8. It was found that the friction coefficient in

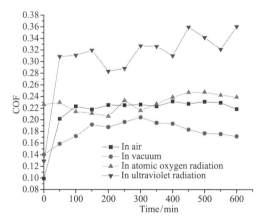

Fig.7.8   The variation of friction coefficient with time for the compound Zn/ZnS layer in four environments

vacuum was the lowest (0.14—0.18). In air environment, it ranged between 0.2 and 0.23. After ultraviolet radiation for 10h, the friction coefficient of the compound Zn/ZnS layer was unstable. The friction coefficient was very low (about 0.13) during the initial stage because of the physical adsorbed layer on the compound layer surface. The adsorbed layer was damaged after 30s when the friction coefficient suddenly increased to 0.3. And then the friction coefficient increased gradually. This phenomenon indicated that the compound Zn/ZnS layer had almost lost lubrication action. Part of the compound Zn/ZnS layer was decomposed, which led to a great fluctuation of the friction coefficient and decrease of lubricating property. The friction coefficient for the composite Zn/ZnS in atomic oxygen radiation was a little higher than that in air, and kept at about 0.22—0.25. It was not stable, because part of the layer was oxidized and the oxide led to the decrease in lubricating property. So it is obvious that the lubricating property of the compound Zn/ZnS layer was impacted by ultraviolet radiation.

7.2.4   Worn Morphologies

Figure 7.9 shows the morphologies of wear scars of the compound Zn/ZnS layer in four environments. It is obvious from Figure 7.9 (a) and (b) that the compound layer was worn more severe in air with obvious wide and deep wear scars. The wear extent was rather large and part of the compound layer flaked off. The wear pattern was mainly adhesion and grain-abrasion wearing. But from the friction coefficient, it was found that the compound layer had not completely lost its lubricating property in air environment, because part of the lubricating film was transferred to the counterpart ball and continued to play the lubricating role. Figure 7.9 (b) shows the morphologies of wear scars of the compound Zn/ZnS layer in vacuum environments. Compared with that in air environment, the wearing was slighter, and the wear track was shallower. Figure 7.9 (c) shows the morphologies of wear scars of the compound Zn/ZnS layer in atomic oxygen radiation. It was obvious that after atomic oxygen radiation for 5h, many big spherical particles with even distribution appeared on the layer surface. The grain boundaries became wide and deep, which was caused by vacuum-pumping. The wearing track was relatively shallow. A part of area was worn out and slight furrows

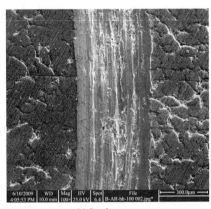
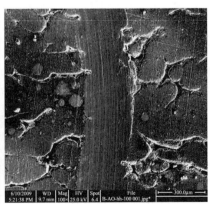

(a) In air   (b) In vacuum

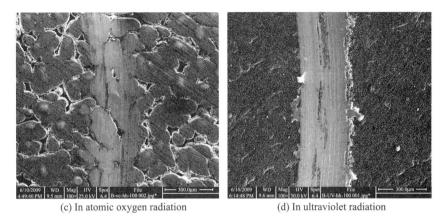

(c) In atomic oxygen radiation   (d) In ultraviolet radiation

Fig.7.9 Worn morphologies of the composite Zn/ZnS layer in four environments

and many micropores occurred. Figure 7.9 (d) shows the morphologies of wear scars of the compound Zn/ZnS layer in ultraviolet radiation environments for 10h. It can be seen that the wear scar of the compound layer in this environment was rougher than that in others. The wearing track was wide and deep, and the layer's surface was worn severely. The layer had been worn out and plastic deformation was observed. The lubricating material accumulated towards both sides of the wearing track to form abrasive dust. So the layer fallen off easily and the adhesive power of layer declined, which led to the decrease in wearing life and lubricating property.

7.2.5   Energy Spectrum Analysis

Figure 7.10 is the EDX analysis result for the compound Zn/ZnS layer in air and vacuum environments. It can be seen from Table 7.2 that oxygen existed in the two environments, and especially the content of oxygen in air was higher (1.55%). This indicates that the surfaces of the layers in the two environments were oxidized slightly during plating, sulfurizing or friction test. From the SEM images, it can be seen that the crystal boundaries were observed clearly, especially in the vacuum environment.

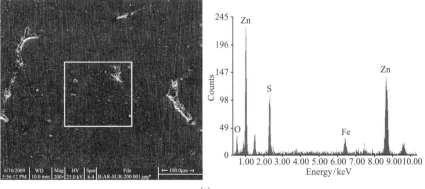

(a)

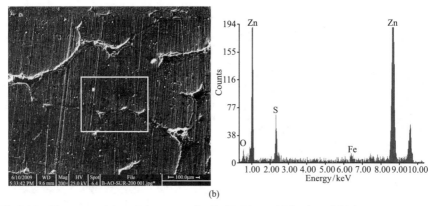

(b)

Fig.7.10   The composition of the composite Zn/ZnS layer (a) in air and (b) in vacuum

Table 7.2   Analysis results of energy spectra (EDAX)

Element	Fe	Zn	S	O
In air/wt%	5.11	89.22	4.12	1.55
In vacuum/wt%	1.08	96.82	1.54	0.56

Figure 7.11 and Table 7.3 are the composition and analysis results of energy spectra for the composite Zn/ZnS layer in atomic oxygen radiation environment. Table 7.3 shows that the oxygen content increased, and it was probably that ZnS had partly decomposed and been oxidized to ZnO. The grain boundaries of the composite layer contribute to the diffusion of high energy oxygen, leading to the oxidization of the layer. Then the adhesive power and lubricating property of the compound layer declined.

Table 7.3   Analysis results of energy spectra (EDAX)

Elements	Fe	Cr	Si	O	S
Particle/wt%	83.81	12.45	1.01	1.43	1.29
	83.68	11.67	0.95	2.13	1.57

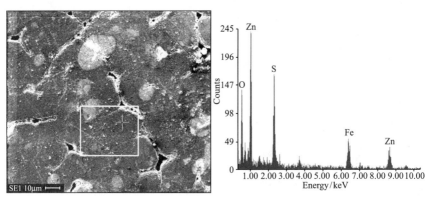

(a)

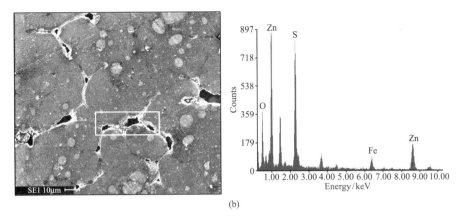

(b)

Fig.7.11   Composition of the composite Zn/ZnS layer in atomic oxygen radiation environment

Figure 7.12 is the EDX analysis result of the white spherical particle on the surface of the compound Zn/ZnS layer in atomic oxygen radiation environment. It can be seen from Table 7.4 that these white spherical particles are mainly Zn, O, S and Fe elements. After atomic oxygen radiation, ZnS was decomposed, and some of Zn was oxidized to ZnO; meanwhile, part of Zn was accumulated to spherical particles.

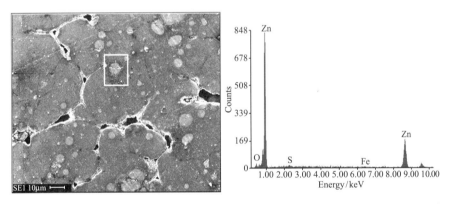

Fig.7.12   The composition of the white particle on the surface of composite Zn/ZnS layer in atomic oxygen radiation environment

Table 7.4   Analysis results of energy spectra (EDAX)

Element	Fe	Zn	S	O
Particle/wt%	00.81	97.74	00.52	00.93

Figure 7.13 and Table 7.5 is the composition and EDX analysis result of the compound Zn/ZnS layer in a random area in ultraviolet radiation environment. Many small particles and a few big particles were present on the compound layer surface and distributed evenly. The components for the small particles are mainly Zn, S, O and Fe.

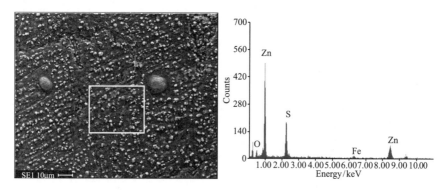

Fig.7.13   The composition of the small particles in a random area on the surface of composite Zn/ZnS layer in ultraviolet radiation environment

Table 7.5   Analysis results of energy spectra (EDAX)

Element	Fe	Zn	S	O
Particle/wt%	3.66	78.87	14.43	3.04

Figure 7.14 is the EDX analysis result of a big particle on the compound Zn/ZnS layer surface in ultraviolet radiation environment. Table 7.6 indicated that the particle was Zn. The data indicated that the energy of ultraviolet wavelength in 100—400nm is enough to cut off the chemical bond. Therefore it can be concluded that part of the ZnS decomposed after ultraviolet radiation, leading to the accumulation of Zn.

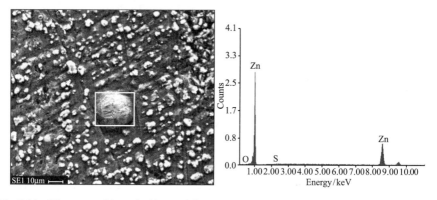

Fig.7.14   The composition of a big particle

Table 7.6   Analysis results of energy spectra (EDAX)

Element	Zn	S	O
Particle/wt%	99.74	0.09	0.16

7.3 Conclusion

The Zn/ZnS composite layer prepared by combined technique-high velocity arc spraying and ion sulfurizing exhibited excellent tribological properties under dry condition at atmosphere, while it showed poor lubrication performance under oil lubrication. The Zn/ZnS composite layer prepared by combined technique-brush plating and ion sulfurizing had excellent anti-friction and wear resistance in vacuum. The lubricating properties were decreased in air and atomic oxygen radiation. However, the lubricating properties were greatly decreased after the composite layer was irradiated by ultraviolet.

Index